Molecular Biology: A Very Short Introduction

VERY SHORT INTRODUCTIONS are for anyone wanting a stimulating and accessible way into a new subject. They are written by experts, and have been translated into more than 45 different languages.

The series began in 1995, and now covers a wide variety of topics in every discipline. The VSI library now contains over 500 volumes—a Very Short Introduction to everything from Psychology and Philosophy of Science to American History and Relativity—and continues to grow in every subject area.

Titles in the series include the following:

Aysha Divan and Janice A. Royds

MOLECULAR BIOLOGY

A Very Short Introduction

OXFORD
UNIVERSITY PRESS

Great Clarendon Street, Oxford, OX2 6DP,
United Kingdom

Oxford University Press is a department of the University of Oxford.
It furthers the University's objective of excellence in research, scholarship,
and education by publishing worldwide. Oxford is a registered trade mark of
Oxford University Press in the UK and in certain other countries

Published in the United States of America by Oxford University Press
198 Madison Avenue, New York, NY 10016, United States of America

British Library Cataloguing in Publication Data
Data available

Library of Congress Control Number: 2016935490

ISBN 978-0-19-872388-2

Printed and bound by
CPI Group (UK) Ltd, Croydon, CR0 4YY

Contents

Preface

This brief account introduces biology at the molecular level. The aim is to provide information and understanding that will enable the reader to enjoy and evaluate the life-changing consequences that have come from the relatively new science of molecular biology.

We journey from the inspirational work of Charles Darwin that gave us the impetus to search for the heritable material to modern-day technology that has enabled us to identify a Plantagenet King in a car park, provide irrefutable forensic evidence, and develop therapeutic targets for new pharmaceuticals. We explore the work that led James Watson and Francis Crick to prove that DNA held the code of life. We look at the work of Alec Jeffreys, who provided forensic tools from 'junk DNA'; so-called because it initially appeared to have no function.

Molecular biology tools are being applied to address global challenges, including sustainable enhancement of food supplies and the improvement of health and well-being. You will learn about cutting-edge technologies that are transforming the life sciences, from high-throughput technologies that allow DNA or proteins to be analysed in a short space of time to synthetic biology and genome editing. These are ushering in an era of more precise medicine, enabling the redesign of existing natural biological systems for medical, agricultural, and other useful purposes and accelerating research and discovery.

While these advancing technologies hold benefits for the public, they also raise concerns about the potential risks to human health, environmental contamination, and deliberate misuse. Dialogue between scientists and the public, legislation, and improved technology are all important in addressing these concerns.

List of illustrations

Common abbreviations

A	adenine
C	cytosine
CDK	cyclin dependent kinase
cDNA	complementary DNA—
DNA	deoxyribonucleic acid
G	guanine
GMO	genetically modified organism
GWAS	genome-wide association studies
HER2	human epidermal growth factor receptor
HR	homologous recombination
IgG	immunoglobulin
IHC	immunohistochemistry
LINE	long interspersed nuclear element
miRNA	micro RNA
mRNA	messenger RNA
mtDNA	mitochondrial DNA
NHEJ	non-homologous end joining
PCR	polymerase chain reaction
pI	isoelectric point
PTM	post-translational modification
qPCR	quantitative PCR
qRT-PCR	quantitative reverse transcriptase PCR
RNA	ribonucleic acid
SINE	short interspersed nuclear element
siRNA	short interfering RNA
SNP	single nucleotide polymorphism
STR	short tandem repeat

T	thymine
TF	transcription factor
TL	telomere length
U	uracil

By convention the names of genes are italicized (e.g. *IGF1*), and the names of proteins are not italicized (e.g. IGF1).

Chapter 1
The early milestones

Molecular biology is the story of the molecules of life, their relationships, and how these interactions are controlled. The early history of molecular biology centred largely on the quest for the nature of the molecules that regulate life, and above all on the identity of the heritable material. Two key ideas in particular provided inspiration at the start of the hunt for the molecular nature of the transmissible factors that provide continuity of life.

First, in the mid-19th century, Charles Darwin and Alfred Russel Wallace both made the case for evolution by natural selection. Darwin presented many examples drawn from his experiences circumnavigating the world on HMS *Beagle*. One of the most notable examples is the thirteen species of Darwin's finches found only on the Galapagos Islands. A single species of finch arriving on these isolated islands approximately two and a half million years ago evolved into thirteen species by natural selection. The food sources differed on each island and the finches adapted the size and shape of their beaks accordingly, sharp and pointed for insects or short and powerful for seeds and nuts. The islands were sufficiently distant from each other to prevent interbreeding, so over time each island developed its own species of finch. Darwin realized that the heritability of advantageous qualities was necessary for natural selection to operate, and result in such adaptations. For an evolutionary radiation of finch species to occur parents must be able to pass stable heritable traits to their offspring. The ability to pass characteristics to subsequent generations is one of the cornerstones of modern molecular biology, but at the time it was in conflict with contemporary ideas that offspring were just a mixture of their parent's traits and that inheritance involved a blending of characteristics. If this were true then evolution would be impeded, if not impossible. Blending would mean that any favourable traits would be lost by dilution through subsequent generations.

The monk and scientist Gregor Mendel, whose work produced the second key idea that led to molecular biology, solved the mechanism of inheritance that had troubled Darwin. While working in his experimental garden at the Abbey of St Thomas in Brno, now in the Czech Republic, Mendel discovered the segregation of heritable traits that led to the birth of genetics. Mendel wanted to study how heritable traits were passed between generations, and he chose to work with pea plants as his model. He studied the transmission of several characteristics, including plant height (short or tall), by the transfer of pollen either naturally or manually. Tall plants that had been bred for several generations always produced tall plants, and likewise short plants were always pure breeders. Mendel discovered that crossing a pure-bred tall plant with a pure small one did not produce a blending of characteristics, but that the resulting progeny were all tall. The tall characteristic must therefore be dominant over the recessive short one. However crossing these new, physically tall plants together produced a mixture of short and tall. The short form did not manifest itself unless two recessive factors came together in an individual. Mendel explained his observations by postulating stable factors 'T' for tall and 't' for short where 'T' is dominant over 't' in appearance. These factors are inherited and passed down the generations (see Figure 1). He did not know about DNA, but from his work he deduced the existence of a heritable factor he called an 'element' and that it must function in inheritance. Wilhelm Johannsen first

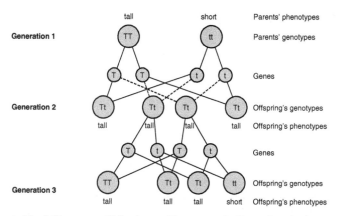

1. Mendel's pattern of inheritance. First cross of tall pea plant (TT) with short pea plant (tt) gives a second generation of all tall-looking (Tt) plants. The second cross using these Tt plants produces tall (TT, Tt) and short (tt) plants.

coined the use of the word 'gene' in 1909 to describe these heritable units. Since then, the heritable factors responsible for a particular trait are known as the genotype and appearance of the trait as the phenotype.

Mendel had resolved the difficulty facing Darwin by demonstrating that the heritable material does not blend as it passes to descendants. This finding was of great significance. It confirms that information for a characteristic is always present in an individual even if it does not show physically in a particular generation. But what was the nature of Mendel's factors or elements? A solution had to wait until 1902 when Theodor Boveri and Walter Sutton independently rediscovered Mendel's work. While working on the development of sea urchins, Boveri, a German biologist, identified chromosomes as the vehicles of heredity that behaved like Mendel's 'factors'. Although in 1888 Heinrich Waldeyer first coined the term 'chromosome' from the Greek for 'coloured body' since they take up dyes easily and become strongly stained, it was Boveri who first realized their

3

significance. He found that somatic or body cells contained two sets of these chromosomes, and sex cells contained only one. The nucleus of a male sperm and of the female egg both contained equivalent amounts of transmissible substance and since each had half the somatic number of chromosomes he reasoned that it was the chromosomes that contained this heritable material. He suggested that chromosomes were dissimilar since a full complement was required for reproduction.

About the same time as Boveri's discovery of chromosomes it was becoming obvious that Mendel's laws applied not just to peas but to all organisms, including humans. Pedigrees from families with inherited diseases provided the first evidence of human genetics. Examples include Huntington's chorea, albinism, Duchenne muscular dystrophy, red–green colour blindness, and haemophilia to name but a few.

The discovery of DNA

Remarkably, both Darwin and Mendel had understood that there must be a heritable material underpinning their observations, but were unaware of its nature. However it was not long after this, in 1869, that Johannes Friedrich Miescher accidentally discovered DNA (deoxyribonucleic acid). Miescher was interested in studying proteins, as these seemed to be the obvious molecules of life carrying out the functions of a cell. He was isolating proteins from white blood cells washed from pus-soaked bandages when he came across a substance that, unlike protein, was resistant to cleavage by protein-digesting enzymes—the proteases—and was also surprisingly high in the chemical phosphate. Miescher in fact thought it was a phosphate storage molecule and he had no idea of its significance. Since it was obviously not protein and it had been found in the cell nucleus, he called it nuclein. In the early 1900s two types of 'nuclein', now called DNA and RNA (ribonucleic acid), were isolated, but at first the differences between them were not apparent. They were named according to their source material, with RNA termed 'Yeast nuclein' and DNA as 'Pancreas nucleic acid'.

At that time nuclein was not considered a suitable candidate for the heritable material and was largely ignored for several decades. Proteins, being more complex, were considered to hold the genetic code. It was almost ninety years after Miescher's discovery that James Watson and Francis Crick in 1953 finally clinched it for DNA with their identification of its double helical structure.

Although Miescher's discovery of DNA was key to the thread of evidence leading to Watson and Crick's momentous discovery, and Sydney Brenner's realization of its key significance, DNA was largely sidelined for many years. It wasn't until 1919 that DNA was analysed to reveal its chemical components, most notably by Phoebus Levene. Levene discovered that DNA was composed of units called nucleotides, and that each nucleotide is made up of three components: a phosphate group, a five-carbon deoxyribose sugar, and a single nitrogen base. These nitrogen bases are of two types called purines and pyrimidines. The purines are adenine (A) and guanine (G), and the pyrimidines cytosine (C) and thymine (T). Each nucleotide is joined to the next through a phosphate group and thus the DNA molecule consists of a string of nucleotides with a sugar-phosphate backbone. The DNA molecule is tightly coiled and packaged into chromosomes. Chromosomes vary in size from the largest human chromosome, known as chromosome 1, and a DNA that extends to 8.5 cm, to the smallest human chromosome 21, with a DNA length of approximately 1.5 cm.

Although the discovery by Watson and Crick in 1953 was hailed by some like Brenner as the dawn of molecular biology, a systematic search for the molecular basis of life had begun in earnest in the 1930s, with the advent of new analysis and separation techniques. The first use of the term 'molecular biology' is attributed to Warren Weaver, the then director of the

Natural Sciences section of the Rockefeller Foundation, who introduced the expression in 1938. Biochemists were seeking to understand the chemistry of biological functions and therefore pioneering separation techniques to isolate the molecules involved. Methods were also developed for the mild disruption of cells in order to release organelles such as nuclei and mitochondria. The molecular components of these organelles could then be separated using techniques that capitalize on differences in the charge or size of the molecules. New compounds were being isolated and their structure and function determined. At the forefront of the biological molecules studied were the three major polymers of life: DNA, RNA, and protein. But how was the genetic information transferred between them? The answer had to wait until the 1960s.

The next key step towards the solution came in the 1940s, from the work by Oswald Avery, Colin MacLeod, and Maclyn McCarty of the Rockefeller Institute, New York. They showed that it was DNA and not protein that could transfer virulence in bacteria and this led them to propose DNA as the heritable molecule. However the scientific community on the whole still needed convincing. Some were hesitant to let go of the idea that only proteins had the complexity to carry the code. In 1952, Alfred Hershey and Martha Chase, working on viruses that infect bacteria called bacteriophages, showed that it was only the DNA from the virus that entered the bacterium and coded for new viral progeny. They concluded that the code was in the DNA and not protein. DNA was becoming increasingly favoured as the molecule mediating the genetic information. After all, DNA resides in the cell nucleus, and the nucleus was the major contribution from the male sperm to a fertilized egg.

There were, however, still some serious doubts that DNA could carry the genetic information. No one could understand how DNA might provide the necessary functions of fidelity for reproduction and sufficient complexity to code for proteins. It was in 1950 that

Erwin Chargaff, an Austrian biochemist who had moved to the USA in 1935 to avoid the Nazi policies in Germany, provided a piece of evidence that would prove crucial. Chargaff found that irrespective of the source, a DNA molecule would always have an amount of adenine equal to that of thymine and similarly cytosine always equalled guanine. Chargaff's rule states that in DNA the larger bases called purines (A+G) exist in a 1:1 ratio with the smaller bases called pyrimidines (C+T), no matter what the species. However the relative amounts of A, C, G, and T vary between individuals and between species, something that would necessarily be the hallmark of the genetic material. Chargaff's rules were central to Watson and Crick's DNA base pair model, although Chargaff himself did not realize the significance of his findings.

From DNA to the Central Dogma

The scene was now set to decipher the structure of DNA and demonstrate how it could fulfil all the functions expected of the genetic material. It was in 1951 that James Watson and Francis Crick started working on the structure of DNA. By this time, there was substantial evidence that DNA was the heritable material but without understanding the structure they could not be sure how it actually worked. The heritable material had to have several properties: faithful duplication on cell division; the ability to hold the code for making proteins; and stability, so that the biological instructions are passed accurately from generation to generation. The structure of DNA therefore must provide an explanation for these known functions, while taking into account all the chemical data known at the time. The X-ray image of Rosalind Franklin's famous photo 51 showed an 'X'-shaped diffraction pattern, consistent with a helical structure for the DNA molecule (see Figure 2).

Chargaff's ratios of A to T and C to G always being equal was at first a puzzle, but using this information in their model-building

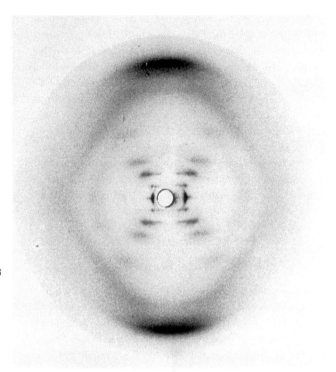

2. Photo 51 showing the X-ray diffraction pattern of double stranded DNA that signifies its helical nature.

set the stage for Watson and Crick's interpretation of DNA structure—as two strands coiled round each other to form a double helix. Significantly, Watson and Crick's model put the bases on the inside of the double helix and the sugar-phosphate groups on the outside, with the two strands connected by weak hydrogen bonds between the bases, paired according to Chargaff's rules. The two strands of a DNA molecule are complementary in sequence; that is, the base A always base pairs with T, and C with G. In the DNA double helix the two strands are antiparallel, which means they run in opposite directions. One strand runs in the 5' to

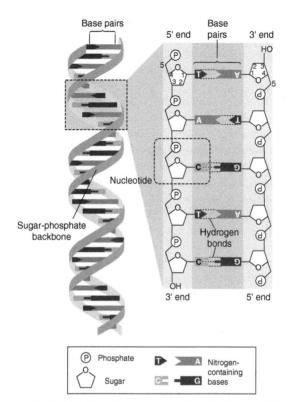

Base pairs

Base pairs

5' end

3' end

HO

5

4   1
3 2

4
3 2 1
O   5

P

P

T ▶ ◀ A

A ◀ ◀ T

P

C ◀ ◀ G

Nucleotide

P

Sugar-phosphate backbone

T ▶ ◀ A

Hydrogen bonds

P

C ◀ ◀ G

OH

3' end

5' end

| P | Phosphate | T ▶ A | Nitrogen-containing bases |
| O | Sugar | G G | |

3. The DNA double helix with a sugar-phosphate backbone held in a double helix by pairing of the bases. The positions of the 5-carbons on the first two deoxyribose sugars are shown.

3' direction (' is pronounced 'prime') and the second strand in the 3' to 5' direction. This nomenclature refers to the carbon atoms in the pentose sugar, numbered one to five. Watson and Crick published their structure in 1953 (see Figure 3).

Once the base-pairing double helical structure of DNA was understood it became apparent that by holding and preserving the genetic code DNA is the source of heredity. The heritable material must also be capable of faithful duplication every time a cell divides. The DNA molecule is ideal for this. It replicates itself by separating its two strands followed by copying each exactly by base-pairing, to give two identical DNA molecules.

The effort then concentrated on how the instructions held by the DNA were translated into the choice of the twenty different amino acids that make up proteins. The Russian-born physicist George Gamow made the suggestion that information held in the four bases of DNA (A, T, C, G) must be read as triplets, called codons. Each codon, made up of three nucleotides, codes for one amino acid or a 'start' or 'stop' signal. This information, which determines an organism's biochemical makeup, is known as the *genetic code.* An encryption based on three nucleotides means that there are sixty-four possible three-letter combinations. But there are only twenty amino acids that are universal. The genetic code is described as redundant or degenerate, as some amino acids can be coded for by more than one codon.

The mechanism of gene expression whereby DNA transfers its information into proteins was determined in the early 1960s by Sydney Brenner, Francois Jacob, and Matthew Meselson. They proposed the existence of an RNA link or messenger between nuclear DNA and the protein synthesis machinery in the cytoplasm. They demonstrated the existence of a messenger RNA by studying a virus infecting a bacterium. The virus makes an RNA from its genome that associates with the host's protein-making facilities to make viral protein. This 'go-between' RNA must be the molecule that carries the code from the DNA to the site of protein synthesis.

Francis Crick proposed in 1958 that information flowed in one direction only: from DNA to RNA to protein. This was called the 'Central Dogma' and describes how DNA is transcribed into RNA, which then acts as a messenger carrying the information to be translated into proteins. Thus the flow of information goes from DNA to RNA to proteins and information can never be

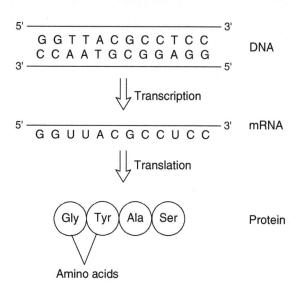

5' ───────────────────────── 3'
        G G T T A C G C C T C C
        C C A A T G C G G A G G
3' ───────────────────────── 5'   DNA

⇓ Transcription

5' ───────────────────────── 3'   mRNA
        G G U U A C G C C U C C

⇓ Translation

(Gly)(Tyr)(Ala)(Ser)   Protein

Amino acids

4. Gene expression: DNA is first transcribed into RNA. RNA is then translated into protein.

transferred back from protein to nucleic acid. DNA can be copied into more DNA (replication) or into RNA (transcription) but only the information in mRNA can be translated into protein (see Figure 4). When in 1961 Sydney Brenner discovered messenger RNA (mRNA), Crick's Central Dogma of DNA, mRNA, and then protein was ratified. Crick however did not preclude the flow of information from RNA to DNA but only from protein back to nucleic acid.

Although molecular biology was not born in 1953 with the discovery of the structure of DNA by Watson and Crick, its elucidation has provided the molecular biologist with tools and techniques that have propelled the science forward. All the information required to make a human being is contained in a single cell. The molecules comprising this fertilized egg will organize the development, sustain the life, allow for the reproduction of, and ultimately execute the demise of an individual. Molecular biology is the study of the way in which molecules function to organize life. Remarkably, the same molecules and principles lie at the heart of all the life sciences, as they control the fundamental machinery of cells. The field of molecular biology concerns macromolecules such as nucleic acids, proteins, carbohydrates, and fats, and their interrelationships, that are essential for life itself. In this chapter, we have completed the first part of the journey through molecular biology: the evolution of ideas and the development of techniques that led to the key finding of the structure of DNA. This groundbreaking discovery set the stage for an explosion in the field of molecular biology.

Chapter 2
DNA

Following the discovery of the structure of DNA, another significant milestone in molecular biology was the publication of the complete sequence of the human genome in 2003. The genome is the entire DNA contained within the forty-six chromosomes located in the nucleus of each human somatic (body) cell. The Human Genome Project, launched in 1990, was an international collaboration involving twenty countries and ranks as perhaps one of the biggest biological experiments undertaken. The aim was to identify the precise order of the bases A, T, G, and C of the full genome by sequencing. The project was coordinated by the National Centre for Human Genome Research in the USA and headed by James Watson, co-discoverer of the structure of DNA. Prior to the project, a low-resolution map of human chromosomes was available identifying where on these chromosomes specific genes were located. However, these were incomplete and not very precise. The whole sequencing process was time-consuming and very costly but enabled a detailed and precise map of the location of genes on the human chromosomes to be built up. It identified many previously unstudied genes and provided a fundamental resource for further research. The publication of the finished sequence coincided with the fiftieth anniversary of the discovery of the DNA helical structure by Watson and Crick amidst much celebration. Soon after, further worldwide projects were launched to work out what the functions of these genes and other regions of the genome actually were.

Components of the human genome

The complete human genome is composed of over 3 billion bases and contains approximately 20,000 genes that code for proteins. This is much lower than earlier estimates of 80,000 to 140,000 and astonished the scientific community when revealed through human genome sequencing. Equally surprising was the finding that genomes of much simpler organisms sequenced at the same time contained a higher number of protein-coding genes than humans. For example, the mustard plant (*Arabidopsis thaliana*), used as a model for studying plant genetics, has a genome size of 125 million bases but the same number of protein-coding genes as the human. It is now clear that the size of a genome does not correspond with the number of protein-coding genes, and these do not determine the complexity of an organism.

Protein-coding genes can be viewed as 'transcription units'. These are made up of sequences called exons that code for amino acids, separated by non-coding sequences called introns. Associated with these are additional sequences termed promoters and enhancers that control the expression of that gene. Genes coding for particular proteins can be single-copy, so that they occur only once in the genome or they can be represented multiple times (multi-copy genes). An additional complexity is that genes can occur as families with each gene in the family coding for similar but non-identical protein members. Gene families can vary in size from two to several hundred and can be confined to a single chromosome or distributed across a number of different chromosomes. As there are two copies of each chromosome in the nucleus of each cell, one set (23) derived from the mother and the second set (23) derived from the father, there are two copies or alleles of each gene. The exception is the XY chromosome pair in the male. As men only have only one X chromosome, they only have one copy of the genes located on chromosome X.

Some sections of the human genome code for RNA molecules that do not have the capacity to produce proteins. A number of these RNAs are involved in the protein synthesis machinery but it is now becoming apparent that many play a role in controlling gene expression. Despite the importance of proteins, less than 1.5 per cent of the genome is made up of exon sequences. A recent estimate is that about 80 per cent of the genome is transcribed or involved in regulatory functions with the rest mainly composed of repetitive sequences.

Satellite DNA, one form of repetitive DNA, is a short sequence repeated many thousands of times in tandem and commonly concentrated at the central region of chromosomes (centromeres). The function of these sequences is unresolved but one role may be to allow correct segregation of chromosomes during cell division. A second type of repetitive DNA is the telomere sequence. These are tandem repeats of the hexanucleotide sequence TTAGGG found at the ends of linear chromosomes. Their role is to prevent chromosomes from shortening during DNA replication and telomere loss is associated with ageing and cancer.

Repetitive sequences can also be found distributed or interspersed throughout the genome. These repeats have the ability to move around the genome and are referred to as mobile or transposable DNA. The most abundant are the long interspersed nuclear elements (LINEs) and the short interspersed nuclear elements (SINEs). Mobile elements can jump from location to location or can make a copy which moves to a new location, leaving the original behind. They are also able to cross over to different locations by combining with repeats of the same sequence elsewhere in the genome, taking intervening sequences with them. Such movements can be harmful sometimes as gene sequences can be disrupted causing disease. For example, one of the causes of Duchenne muscular dystrophy is insertion of the LINE 1 element within the dystrophin gene. This results in the loss of the fully functional muscular protein dystrophin and leads to the progressive muscle weakness observed in these patients. The vast majority of transposable sequences are no longer able to move around and are considered to be 'silent'. However, these movements have contributed, over evolutionary time, to the organization and evolution of the genome, by creating new or modified genes leading to the production of proteins with novel functions. The creation of gene families is a good example of this.

Organization of the human genome

The total length of DNA in each cell varies by species but can be up to several metres long. For example, the DNA in each human cell is roughly two metres in length. This linear molecule has to be packaged so that it can fit into the cell nucleus, the diameter of which is approximately one million times smaller (6 microns). To achieve this, the DNA is packaged at several levels. At the first level linear DNA is attached to proteins called histones to form structures called nucleosomes. Each nucleosome is made up of 146 base pairs of DNA wrapped around eight histone protein molecules and separated from the next nucleosome by short linker DNA. These appear as a 'beads on a string' structure. This chain of nucleosomes folds up to produce a fibre, which is then further compacted into an array of large loops. These loops are then coiled to produce the chromosome. This highly ordered DNA-protein structure that makes up the eukaryotic chromosomes is called chromatin (see Figure 5).

Chromatin packaging is not uniform across the chromosomes. Some regions are less compact than others. Loose chromatin, termed euchromatin, is associated with regions of the DNA that are replicating, or with genes being transcribed. The more open structure allows enzymes involved in these processes to access the chromosomal regions. Conversely, compact chromatin, termed

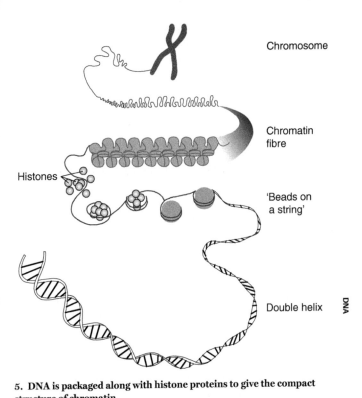

Chromosome

Chromatin fibre

Histones

'Beads on a string'

Double helix

5. DNA is packaged along with histone proteins to give the compact structure of chromatin.

heterochromatin, is associated with regions of the DNA that are not replicating or are transcriptionally inactive.

Not all DNA is packaged within the nucleus. In humans, a small amount also resides in the mitochondria, termed mitochondrial DNA (mtDNA). Mitochondria are structures within the cells that convert energy stored in food into a form that cells can use to power their processes. mtDNA is made up of around 16,500 base pairs, coding for 37 genes whose protein products are involved in energy production. mtDNA is used in tracing ancestry; as it is passed from mother to offspring, relatives within the same maternal lineage share the same mtDNA. mtDNA is also used as an identifier in genetic fingerprinting in forensic science.

DNA replication

A very important property of DNA is that it can make an accurate copy of itself. This is necessary since cells die during the normal wear and tear of tissues and need to be replenished. Cells are replenished through the cell division cycle, during which a cell replicates its DNA to produce two identical copies, which are then separated into two daughter cells.

During replication, the DNA double helix is first unwound from the nucleosomes and the two strands are separated to expose unpaired bases. This unwinding is carried out by a protein called helicase and occurs at sites called origins of replication. Each exposed strand then acts as a template for the synthesis of new DNA. This reaction is carried out by a group of enzymes called DNA polymerases. To start synthesis, DNA polymerase requires a short RNA primer consisting of about ten nucleotides. Free nucleotides are added to the end of this primer matched to the template strand by complementary base-pairing. DNA polymerase can only extend the nucleotide chain in one direction, from the 5' to the 3' end, and so the antiparallel DNA strands are synthesized slightly differently. One strand, the leading strand, is synthesized as a continuous piece of DNA using a single RNA primer. The second strand, the lagging strand, uses several RNA primers and is synthesized as short fragments called Okazaki fragments after the Japanese scientist Reiji Okazaki, who discovered them (see Figure 6). Once replication is complete, the RNA primers are removed, the gaps filled by DNA polymerase, and any fragments joined by the enzyme DNA ligase. The DNA then zips up and rewinds to form the DNA double helix.

At the end of replication, each DNA molecule contains one template DNA strand and one newly synthesized strand. This is

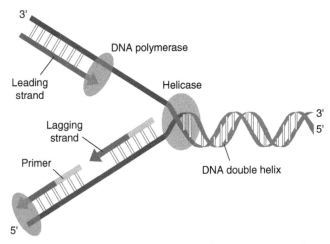

6. DNA is synthesized by the enzyme DNA polymerase; one strand as a leading (continuous) strand and the second as a lagging (discontinuous) strand.

called semi-conservative replication as half of the original double-stranded DNA molecule is conserved in each of the daughter molecules. At the end of the process when the cells divide into two, each new cell acquires an exact copy of the DNA present in the original cell.

Mutations and correction mechanisms

DNA replication is a highly accurate process with an error occurring every 10,000 to 1 million bases in human DNA. This low frequency is because the DNA polymerases carry a proofreading function. If an incorrect nucleotide is incorporated during DNA synthesis, the polymerase detects the error and excises the incorrect base. Following excision, the polymerase reinserts the correct base and replication continues. Any errors that are not corrected through proofreading are repaired by an alternative mismatch repair mechanism.

In some instances, proofreading and repair mechanisms fail to correct errors. These become permanent mutations after the next cell division cycle as they are no longer recognized as errors and are therefore propagated each time the DNA replicates. Mutations are of several types; point mutations, in which a single nucleotide is substituted for another, or insertion or deletion mutations, in which one or more nucleotides are added or deleted. Sometimes, mutations are silent, in that no phenotypic change occurs. However, mutations can cause disease, if the normal function of the protein is lost or the protein acquires new but pathogenic (harmful) functions. In both cases, the cellular processes in which the protein acts are disrupted.

Sickle cell anaemia is a disease caused by a point mutation in the gene coding for the beta-globin protein of haemoglobin, the oxygen-carrying molecule found in red blood cells. The mutation causes the hydrophilic (water-loving) amino acid glutamic acid to be replaced by the hydrophobic (water-hating) amino acid valine at the sixth position of beta-globin. An altered protein is produced that causes the haemoglobin to aggregate, distorting the red blood cell to a sickle shape. These sickle-shaped red blood cells do not flow as easily through the bloodstream and can block blood vessels, resulting in tissue or organ damage.

Some mutations can confer an advantage to an organism, as is observed in the example of sickle cell anaemia. Individuals who have a single mutated sickle cell allele instead of two mutated alleles are carriers of the disease and do not show disease symptoms of the same severity. These carriers are resistant to the effects of malaria and hence sickle cell alleles are most common in individuals from parts of the world where the incidence of malaria is high.

Polymorphisms

The DNA sequence between humans is 99.5 per cent identical and it is the remaining 0.5 per cent which provides the diversity we see between individuals. Mutations are one way in which this genetic variation arises, creating alternative forms of DNA, termed polymorphisms. The most common types are the Single Nucleotide Polymorphisms or SNPs (pronounced 'snips'). At such sites, the genomes of individuals differ by a single nucleotide. To be classified as an SNP, the variation must occur in more than 1 per cent of the population. SNPs were first found in 1978 in the beta-globin gene cluster and since then approximately 10 million have been identified. They can arise in coding or non-coding regions of the DNA and most have no effect on health or development. SNPs can, however, be associated with certain disorders and these can be used as markers to identify regions of the genome where a disease-causing gene or genes are located. In genome-wide association studies (GWAS), genome sequences of a large group of people affected by a disease are compared with a second large unaffected group. If certain SNPs are found to occur more frequently in people with the disease then these are said to be associated with the disease. Further analysis of that particular region of the genome can then be undertaken to identify the exact genetic change involved in the disease. One of the early successes of GWAS was published in 2005. The authors identified a variation in the complement factor H gene as a risk factor for age-related macular degeneration, a disease that leads to visual impairment and blindness in the elderly. Since then, GWAS has been used to identify genetic variations that contribute to many disorders including cancers, Alzheimer's, diabetes, and heart disease. The National Human Genome Research Institute (NHGRI) and the European Bioinformatics Institute (EBI) maintain catalogues of published GWAS data that is freely available and can be used to search for SNPs associated with common diseases or traits.

How do we study DNA?

Advances in DNA research have only been possible due to the development of DNA technologies in the past fifty years, including gene cloning, the polymerase chain reaction (PCR), and sequencing methods.

Gene cloning

Gene cloning refers to a set of techniques that enable us to make many identical copies of regions of DNA. This is not to be confused with whole organism cloning like that of Dolly the Sheep. In the latter, known as reproductive cloning, an exact copy of the entire organism is produced that has the same nuclear DNA as an existing animal. In gene cloning, smaller regions of DNA are copied for analysis in the laboratory or for other applications. These techniques enable us to isolate DNA sequences of interest and produce them in large quantities so that their function can be studied. It also enables us to manipulate sequences to produce molecules with new functions for use in medicine and agriculture—areas that will be covered in Chapters 6 and 9.

In gene cloning experiments, the first step is to generate recombinant DNA. This involves inserting a DNA fragment of interest into a self-replicating molecule called a vector. One that is widely used is the bacterial plasmid. This is a circular double-stranded DNA molecule which is physically separate from chromosomal DNA. It is capable of self-replicating as it carries its own origin of replication.

To insert the DNA fragment into the vector, both are first cut by special enzymes called restriction endonucleases. These generate breaks at specific sequences in the DNA in a way that enables molecules cut with the same enzyme to be joined together using DNA ligase (see Figure 7). Construction of recombinant DNA molecules was only possible after the discovery of these two enzymes; DNA ligase in 1967 and restriction enzymes in 1970. Scientists were now able to cut DNA molecules at specific locations and recombine the fragments in different combinations. To facilitate recombinant DNA generation, vectors are engineered to carry a multiple cloning site (MCS). This region contains several

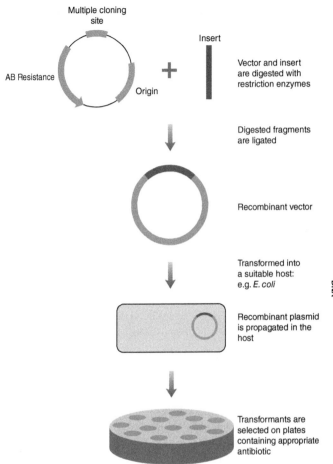

7. Steps in a gene cloning experiment. DNA (insert) is ligated into a vector, transformed into a host cell, and plated on agar.

restriction enzyme recognition sites positioned very close to each other. MCSs provide flexibility in cloning experiments by extending the choice of restriction endonucleases that can be used by researchers.

Once recombinant DNA is generated, it is transferred into a host cell where it is replicated to make many identical copies or clones. A cell that takes up DNA is called a transformed cell. Host cells can be bacterial cells, yeast cells, animal or plant cells. Generally the DNA to be cloned, termed the target or source DNA, is derived from a different species from that of the host and thus introduction of the source DNA into the host results in the production of a transgenic organism or a genetically modified organism (GMO).

Recombinant DNA molecules are introduced into the host typically by treating the cells with chemicals and then applying a brief heat shock or an electric current. This process makes the cells porous and facilitates uptake of recombinant DNA. To identify cells that have been transformed, a selection system is required. This is achieved by engineering vectors to carry marker genes, usually an antibiotic resistance gene which the host is sensitive to. When cells are plated out onto agar plates containing the antibiotic, cells that have taken up the vector grow, while those that have not taken up the vector die. As the transformed cells grow and divide, clones of cells (colonies) appear on these plates.

Cloning experiments require careful planning if they are to be successful. An important part of this is selecting an appropriate vector and then matching it to a host system in which it will function. Vectors can be purchased commercially and a large number are available derived from bacterial, viral, or yeast sources. All are engineered to carry three basic features; an MCS into which the DNA fragment can be inserted, an origin of replication to enable copying within the host cell, and a marker gene for recombinant clone selection. For whole genome sequencing experiments, vectors called Bacterial Artificial Chromosomes (BACs) are used as these allow cloning of large DNA fragments and can be propagated in bacterial host cells. Viral vectors are often used if the host cell is mammalian as they can be introduced more easily into these cells.

Electrophoresis

It is often necessary in DNA experiments to identify specific fragments of DNA from a mixture. For example, in many cloning experiments, when DNA is cut using restriction enzymes, a series of smaller-sized fragments are generated. These can be separated on the basis of size by a technique developed in the 1970s called gel electrophoresis. The DNA mixture is loaded onto a gel made from agarose or polyacrylamide. As the gels are porous, when an electric current is applied, DNA travels through the gel with the smaller molecules moving more rapidly than the larger ones. The gel is then stained with special dyes to visualize the DNA. The DNA fragments appear as bands revealing the position of the different sizes. In 1975, Edwin Southern showed that it was possible to transfer, or blot, the DNA fragments from a gel onto a membrane. A specific DNA fragment of interest could then be located by adding a probe to the membrane. Probes are short sequences designed to be complementary to the DNA sequence being studied and when applied to the membrane, bind or hybridize to its target. As a fluorescent, chemical, or radioactive label is attached to the probe, the hybridized target is detected as a sharp band when exposed to specific conditions. This technique, named Southern blotting after its inventor Edwin Southern, revolutionized how DNA was studied. Up until that time, genomic DNA, digested with restriction enzymes and separated by electrophoresis, produced a smear of bands. It was often difficult to identify specific fragments from this smear. With Southern blotting, it was now possible to identify a particular gene within an entire genome and this facilitated the discovery of genetic defects such as sickle cell anaemia. Southern blotting has now largely been superseded by more rapid and sensitive techniques like PCR and sequencing. However, gel electrophoresis continues to be used regularly in molecular biology laboratories.

The polymerase chain reaction

The polymerase chain reaction (PCR) is a very powerful laboratory technique that utilizes the natural process of DNA replication. It allows amplification of sections of DNA to produce large quantities from only small amounts of starter DNA. The method was developed in the 1980s by Kary Mullis, for which he was awarded a Nobel Prize in 1993. PCR is used in many areas of science, in DNA cloning experiments, establishing paternity or other biological relationships, diagnosing genetic and infectious diseases, and in forensic studies.

Widespread use of this technique was facilitated by the discovery of the thermostable enzyme *Taq* polymerase. This DNA polymerase was initially isolated from the bacterium *Thermus aquaticus* found in geothermal hot springs. It can withstand the high temperatures required to separate the DNA strands in the laboratory and thus Mullis used it in PCR. Prior to this a thermo-labile polymerase was used. However, as PCR requires repeated cycles of heating and cooling, this polymerase was inactivated and so after each high temperature step, the reaction had to be stopped to add fresh enzyme. Introduction of *Taq* polymerase led to PCR becoming automated using thermal-cycling machines. The process can now be completed in a few hours compared to the laborious stop-start reactions performed in kettles in the early days.

To carry out PCR, the DNA to be amplified (the template) is added to a tube, along with DNA polymerase, nucleotides, primers, and chemical reagents which provide optimal conditions for the reaction to occur. The template can be genomic DNA extracted from humans, microbes, or plants. To define the region to be amplified, primers are designed flanking the sequence of interest. Amplification is achieved through successive rounds of denaturation, annealing, and extension with each step occurring at a different temperature (see Figure 8). During denaturation, the DNA double-strands are separated (denatured) typically by heating to 94°C. The temperature is then lowered to around 55°C at which point the primers anneal to their complementary regions on opposite strands of the single-stranded DNA. In the third step, the temperature is raised typically to 72°C, which enables the

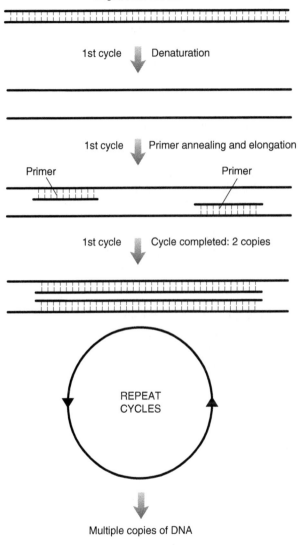

Original DNA strand

1st cycle ▼ Denaturation

1st cycle ▼ Primer annealing and elongation

Primer   Primer

1st cycle ▼ Cycle completed: 2 copies

DNA

REPEAT CYCLES

Multiple copies of DNA

8. The polymerase chain reaction. Amplification of target sequence is achieved through successive rounds of denaturation, annealing, and elongation.

DNA polymerase to extend the primers by adding nucleotides complementary to the template. At the end of these three steps, an identical copy of the original DNA is made. The cycle then begins again using both the original and the newly synthesized DNA as a template and is repeated thirty to forty times. At the end of each cycle, the number of copies of the template DNA is increased exponentially. One DNA molecule produces two copies, then four, then eight, and so forth so that after thirty or forty cycles, billions of copies of the original DNA are made.

Many variations of the traditional PCR technique have now been developed. One is Reverse Transcriptase PCR (RT-PCR) in which RNA is used as the starter material. The RNA is first converted to complementary DNA (cDNA) using an enzyme called reverse transcriptase. Subsequently the cDNA is amplified using cycles of denaturation, annealing, and extension as in the conventional PCR technique. cDNA is identical to the DNA sequence from which the RNA is made but does not contain introns or other genome elements such as promoter regions. cDNA is often used in gene cloning experiments where protein production in a host cell is required. Another variation is Quantitative PCR (qPCR) in which the target DNA or RNA is amplified and quantified simultaneously. This is particularly useful in research but also to diagnose genetic diseases by identifying SNPs or other changes such as gene copy number variations.

DNA sequencing

DNA sequencing identifies the precise linear order of the nucleotide bases A, C, G, T, in a DNA fragment. It is possible to sequence individual genes, segments of a genome, or whole genomes. Sequencing information is fundamental in helping us understand how our genome is structured and how it functions. It has been used to build up genome maps of different life forms, including humans, animals, and microbes. It can help identify genes responsible for causing inherited genetic diseases like cystic fibrosis but also other diseases where gene function is impaired such as cancer and diabetes. By comparing the genomes of different organisms, sequencing can provide clues as to which genes are conserved between species and thus are likely to be functionally important. It can also help catalogue and classify different species which can then be used, for example, to identify disease-causing pathogens or contamination of food.

The first method widely used to sequence DNA was the Sanger Sequencing method developed by Frederick Sanger in 1977 for which he was awarded a Nobel Prize in 1980. In this method the DNA to be sequenced is copied repeatedly by DNA polymerase by inserting complementary nucleotides into the copied strand. However, the nucleotides added are modified chemically so that when the modified nucleotide is incorporated into the growing chain, the copying reactions stop. This nucleotide has a fluorescent tag attached to it and therefore the process generates a set of fragments which differ in size by one nucleotide ending with a fluorescent label. These fragments are separated by size by electrophoresis and the sequence read by identifying the end fluorescent nucleotide using an automated sequencing machine. From this the original DNA sequence can be recreated (see Figure 9).

Although Sanger Sequencing is still used, it is now increasingly being replaced by newer technologies that are developing at an astounding pace. These technologies, collectively referred to as next-generation or high-throughput sequencing, allow DNA to be sequenced much more quickly and cheaply. The Human Genome Project, which used Sanger sequencing, took ten years to sequence and cost 3 billion US dollars. Using high-throughput sequencing, the entire human genome can now be sequenced in a few days at a cost of 3,000 US dollars. These costs are continuing to fall, making it more feasible to sequence whole genomes.

The human genome sequence published in 2003 was built from DNA pooled from a number of donors to generate a 'reference' or

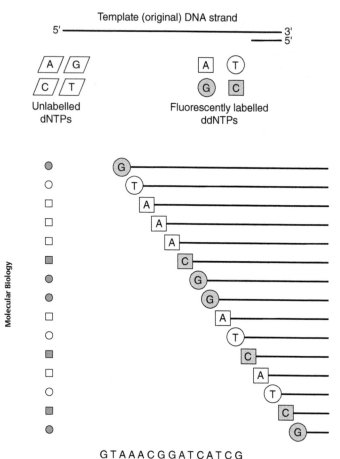

GTAAACGGATCATCG

9. The Sanger sequencing method. Fragments generated during the process are separated by size and the sequence read by identifying the end fluorescent nucleotide.

composite genome. However, the genome of each individual is unique and so in 2005 the Personal Genome Project was launched in the USA aiming to sequence and analyse the genomes of 100,000 volunteers across the world. Soon after, similar projects followed in Canada and Korea and, in 2013, in the UK. A number of well-known individuals have had their genomes sequenced, including James Watson and Steve Jobs, co-founder of Apple Inc. All that is required of volunteers is a biological sample from which the DNA can be extracted and sequenced. As the Personal Genome Project is intended to help scientists learn more about how genes interact with the environment to cause disease, details about the individual's phenotype, health, and lifestyle are also collected. Personal genomics can be used to identify gene variants that may increase an individual's risk of developing particular diseases. This could allow doctors to intervene with risk-reducing medication or procedures or motivate individuals to make lifestyle changes. One of the first patients to be diagnosed through the UK-based Personal Genomes Project was the Hedley family in 2015. Mr Hedley, aged 57, had a lifelong history of high blood pressure leading to kidney failure. His brother, father, and uncle had all died of kidney failure and his daughter aged 34 was also showing early signs of kidney damage. Genome sequencing revealed that this rare kidney disease is caused by a specific genetic variant carried by both father and daughter but not the granddaughter. Doctors are now treating the affected family members with existing drugs to control the disease.

Although personal genomics promises improved health and well-being, the challenges are substantial, including the storage and interpretation of the overwhelmingly large amounts of data that is generated. As one of the goals of the Personal Genome Project is to make the sequencing data openly accessible through databases this also raises issues around privacy and how the information could be used by others.

Bioinformatics

To store and analyse the huge amounts of data, computational systems have developed in parallel. This branch of biology, called bioinformatics, has become an extremely important collaborative research area for molecular biologists drawing on the expertise of computer scientists, mathematicians, and statisticians. DNA sequences are routinely deposited into databases that are freely available to the public. The best-known DNA sequence databases are Genbank at NCBI (National Center for Biotechnology Information) and the EMBL (European Molecular Biology Laboratory) Nucleotide Sequence Database maintained by EBI. Making sequencing data freely available was a goal of the Human Genome Project and has continued since to facilitate research through sharing of information. Other biological databases also exist, in which protein sequences and data from studies involving RNA are stored. By selecting an appropriate computer program, scientists can use the data to search for genes, identify gene function, and work with three-dimensional models of protein structure.

Chapter 3
RNA

The discovery of RNA was made in the early 1900s but the difference between DNA and RNA was not apparent at that time. However, by the mid-1900s it was clear that RNA was distinctly different from DNA in structure and function. The first RNA molecules to be discovered were those involved in protein synthesis, mRNA, transfer RNA (tRNA), and ribosomal RNA (rRNA). In recent years, a vast number of additional RNA molecules have been identified. These are non-coding RNAs involved not in protein synthesis but in regulating gene expression. These discoveries have been facilitated by large-scale international projects such as ENCODE (Encyclopaedia of DNA elements) launched in 2003, which aims to characterize the functions, if any, of non-protein-coding components of the genome. With new non-coding RNAs being discovered at a breathtaking pace it is now becoming apparent that the RNA world is much richer than we first suspected.

RNA, like DNA, is a chain composed of repeating nucleotides, with each nucleotide linked to the next through a chemical bond. However, the structure of RNA differs from DNA in three fundamental ways. First, the sugar is a ribose, whereas in DNA it is a deoxyribose. Secondly, in RNA the nucleotide bases are A, G, C, and U (uracil) instead of A, G, C, and T. U and T have similar base-pairing properties and thus uracil base pairs with adenine.

Thirdly, RNA is a single-stranded molecule unlike double-stranded DNA. It is not helical in shape but can fold to form a hairpin or stem-loop structure by base-pairing between complementary regions within the same RNA molecule. These two-dimensional secondary structures can further fold to form complex three-dimensional, tertiary structures. An RNA molecule is able to interact not only with itself, but also with other RNAs, with DNA, and with proteins. These interactions, and the variety of conformations that RNAs can adopt, enables them to carry out a wide range of functions.

RNAs in protein synthesis

Proteins are synthesized in the cell by a two-step process, transcription and translation. During transcription, the gene coding for the protein to be produced is first transcribed into mRNA that is subsequently translated into one or more proteins.

Transcription

In transcription, the mRNA is produced in the nucleus of the cell by the enzyme RNA polymerase II, using one strand of the DNA double helix as a template. This template strand is called the non-coding strand. Its sequence is complementary to the code so that the mRNA transcribed from it will carry the coding sequence, enabling it to direct protein synthesis. Transcription is initiated when a number of proteins assemble in a defined order at a specific site on the DNA called the promoter. The promoter is usually located 25–35 base pairs away, upstream from the transcription start site—that is, the point from which mRNA synthesis will start. The best-defined promoter sequence in eukaryotes is the TATA box which is recognized by a protein called Transcription Factor IID (TFIID). Once TFIID binds, RNA polymerase II and other proteins also combine, and synthesis of the mRNA transcript begins. RNA polymerase II moves along the DNA template, adding complementary bases to extend the RNA strand. Transcription is terminated when RNA polymerase II

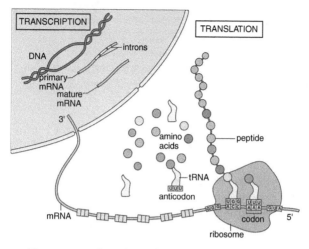

10. The two steps of protein synthesis; transcription in the nucleus and translation at the ribosomes in the cytoplasm.

encounters a stop codon (TAA, TAG, or TGA) at the end of the transcription unit.

Transcription produces a pre-mRNA transcript which is modified to form mature mRNA before translation into a protein sequence. The transcript is capped at the 5' end by the addition of a modified guanine nucleotide and a poly-A tail of around 250 adenine nucleotides is added at the 3' end. The cap and tail prevents the mRNA from being degraded and allows it to pass from the nucleus to the cytoplasm, where translation takes place (see Figure 10).

The mRNA transcript like DNA contains intron and exon sequences. Prior to the mRNA passing to the cytoplasm, the intervening introns are removed—spliced out—and adjacent coding exons are joined back together. These are carried out by a class of non-coding RNAs called small nuclear RNAs (snRNA). From these modifications a mature transcript made up of exons, a

5′ cap, and a 3′ tail is produced. It also contains sequences at the end of the transcript preceding the cap and the tail called untranslated regions or UTR sequences. These UTRs contain regulatory information such as how frequently the transcript will be translated into protein and how long it will survive in the cell before being degraded.

Translation

Translation of the mRNA to a protein takes place in the cell cytoplasm on ribosomes. Ribosomes are cellular structures made up primarily of rRNA and proteins. At the ribosomes, the mRNA is decoded to produce a specific protein according to the rules defined by the genetic code. The correct amino acids are brought to the mRNA at the ribosomes by molecules called transfer RNAs (tRNAs). These contain a three-nucleotide sequence that is complementary to the codon on the mRNA and carries an amino acid that corresponds to that sequence. At the start of translation, a tRNA binds to the mRNA at the start codon AUG. This is followed by the binding of a second tRNA matching the adjacent mRNA codon. The two neighbouring amino acids linked to the tRNAs are joined together by a chemical bond called the peptide bond. Once the peptide bond forms, the first tRNA detaches leaving its amino acid behind. The ribosome then moves one codon along the mRNA and a third tRNA binds. In this way, tRNAs sequentially bind to the mRNA as the ribosome moves from codon to codon. Each time a tRNA molecule binds, the linked amino acid is transferred to the growing amino acid chain. Thus the mRNA sequence is translated into a chain of amino acids connected by peptide bonds to produce a polypeptide chain. Translation is terminated when the ribosome encounters a stop codon (UAA, UAG, or UGA). At this point the ribosome releases the finished polypeptide chain. After translation, the chain is folded and very often modified by the addition of sugar or other molecules to produce fully functional proteins.

Regulatory RNAs

In the last two decades it has become apparent that RNAs are involved in much more than simply acting as a messenger between DNA and protein, forming structural components of ribosomes, and shuttling amino acids to the ribosomes. We now know that RNAs can influence many normal cellular and disease processes by regulating gene expression. RNA interference, abbreviated to RNAi, is one of the main ways in which gene expression is regulated. In this process, a short double-stranded RNA, complementary in sequence to a specific mRNA sequence, binds to it and prevents the mRNA from producing a protein. Thus the corresponding gene is 'silenced'.

RNAi was discovered unexpectedly by two researchers, Andrew Fire and Craig Mello, who were trying to manipulate the expression of specific genes in the nematode *C. elegans*. They found that injecting double-stranded RNA into the nematode silenced the expression of the target gene much more effectively than adding single-stranded RNA alone. Since the publication of Fire and Mello's discovery in 1998, RNAi has been extensively studied within several organisms including the fruit fly, yeast, plants, and humans. Three classes of RNAi molecules have been identified to date with micro RNAs (miRNA) being one of the main regulators of gene expression in humans.

miRNAs

miRNAs are transcribed from several different loci in the human genome. They are produced as long RNAs which fold to form a double-stranded hairpin structure. The presence of double-stranded RNA within a cell activates the RNAi machinery which ultimately leads to gene silencing. Around 1,000 genes coding for miRNAs in humans have been identified, regulating a whole range of different biological processes at all stages of human development, from embryo through to adulthood. Their dysfunction is associated with a range of diseases including cancers, heart disease, and immunological disorders—examples of which will be covered in Chapters 5, 7, and 9.

Gene silencing through the RNAi pathway is a complex process. The presence of double-stranded RNA molecules in a cell triggers the activation of an enzyme called DICER. DICER cuts the long strands into smaller fragments, approximately 22 nucleotides in length, to produce short interfering RNAs (siRNA). One of the two strands of siRNA is then loaded onto a multi-protein complex called RISC (RNA-induced silencing complex). The second strand is discarded. RISC directs the loaded strand to bind to its complementary mRNA. Once bound, the enzyme Argonaute, a component of RISC, cleaves the mRNA molecule or prevents translation from occurring. This silences the expression of the gene and the protein encoded by that mRNA is not produced (see Figure 11).

RNAi in research and in therapy

The naturally occurring RNAi pathway is now extensively exploited in the laboratory to study the function of genes. It is possible to design synthetic siRNA molecules with a sequence complementary to the gene under study. These double-stranded RNA molecules are then introduced into the cell by special techniques to temporarily knock down the expression of that gene. By studying the phenotypic effects of this severe reduction of gene expression, the function of that gene can be identified.

Synthetic siRNA molecules also have the potential to be used to treat diseases. If a disease is caused or enhanced by a particular gene product, then siRNAs can be designed against that gene to silence its expression. This prevents the protein which drives the disease from being produced. For example, familial hypercholesterolemia (FH) is a genetic disorder characterized by high levels of cholesterol and increased risk of coronary heart

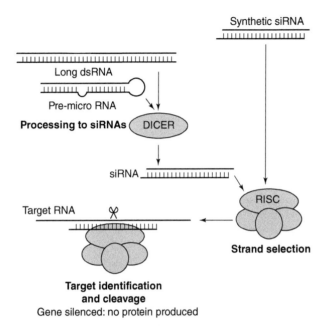

Synthetic siRNA

Long dsRNA

Pre-micro RNA

Processing to siRNAs (DICER)

siRNA

Target RNA

RISC

Strand selection

**Target identification
and cleavage**
Gene silenced: no protein produced

**11. The RNAi pathway. siRNA silences gene expression by binding to
its target mRNA and preventing protein production.**

disease. One of the mutations responsible for this condition is
located in the gene *PCSK9* which codes for an enzyme that raises
the level of low-density lipoprotein (LDL)—a bad form of
cholesterol. Blocking the production of PCSK9 using siRNA
would result in reduction of LDL and hence could be used to treat
FH. This approach has shown to be effective in early clinical trials.

siRNA therapy is also being trialled with a number of other
conditions including cancers, liver disease, and viral infections.
One of the major challenges to the use of RNAi as therapy is
directing siRNA to the specific cells in which gene silencing is
required. If released directly into the bloodstream, enzymes in the
bloodstream degrade siRNAs. In addition, they cannot pass

through the hydrophobic cell membrane and enter the cell since they are negatively charged molecules. Other problems are that siRNAs can stimulate the body's immune response and can produce off-target effects by silencing RNA molecules other than those against which they are specifically designed. Some of the earliest clinical trials involved injecting siRNAs directly into the eye to treat vision loss in age-related macular degeneration and diabetic macular oedema. However, direct injections are not possible for all conditions and considerable attention is currently focused on designing carrier molecules that can transport siRNA through the bloodstream to the diseased cell. One carrier molecule is the lipid-based nanoparticles (LNP). These are very small molecules, 70–80 nm in size, into which the siRNA is packaged. siRNAs can thus be transported through the bloodstream, protected from the degrading action of enzymes, and can also pass through the hydrophobic cell membrane into the cell cytoplasm. Some of these carriers co-carry targeting molecules so that the siRNA is taken up specifically by the diseased cell it is targeting. Calando Pharmaceuticals first tested this strategy by coating the nanoparticles carrying siRNA with a protein called transferrin. Transferrin binds to a specific protein called the transferrin receptor found on the surface of cells. Once bound, the nanoparticle is internalized entering the cell cytoplasm. The transferrin receptor is found in high amounts on the surface of tumour cells compared to non-tumour cells and thus tumour cells preferentially take up the nanoparticle. This approach is also being trialled to treat patients with other diseases such as those infected with the hepatitis B virus.

Long non-coding RNAs

Another class of non-coding RNAs recently identified are the long non-coding RNAs (lncRNA). These are RNA molecules which are typically transcribed from regions of the genome that lie between transcription units. However, they can also be transcribed from within the exons or introns of protein-coding genes. LncRNAs are distinguished from the smaller miRNAs and siRNAs by their length. LncRNAs are greater than 200 nucleotides in length, compared to the 21- to 35-nucleotide length of the smaller RNAs.

Approximately 10,000 lncRNAs have been identified in humans so far through the ENCODE project. The functions of the majority of these have not yet been characterized. However, of those studied, it is evident that lncRNAs play a role in regulating the expression of protein-coding genes at many levels. They can control whether a particular gene is transcribed or not, the splicing of mRNA transcripts, and whether the mRNA transcript is translated to protein or not. LncRNAs have also been implicated in the development of a number of diseases, including cancers, neurological disorders, immune-mediated diseases, and cardiovascular diseases. Some pathways in which lncRNAs act will be covered in Chapter 5.

Catalytic RNAs

Yet another class of RNAs are the ribozymes, first discovered in 1982 in bacteria and subsequently in eukaryotic organisms. These RNA molecules act as enzymes catalysing specific biochemical reactions in a similar way to that of protein enzymes. Consequently, they are also known as catalytic RNAs. Some ribozymes, such as the snRNAs, catalyse splicing of pre-mRNA transcripts to cut out the intervening intron sequences and join the adjacent exon sequences together. rRNA, the RNA component of ribosomes, is also a catalytic RNA. It catalyses the formation of peptide bonds linking amino acids together in a polypeptide chain during protein synthesis.

Ribozymes can be constructed artificially in the laboratory for therapeutic purposes. Here a ribozyme is engineered to bind to and cleave a target mRNA, thereby knocking down the expression of that gene. Hence, it uses a similar approach to treatment as siRNA by silencing the expression of disease-causing or -enhancing genes. Ribozymes targeting specific RNAs have been tested in a number of clinical trials as treatment for cancers such as kidney and breast and some viral infections including Human Immunodeficiency Virus (HIV) and hepatitis C virus. Ribozyme-based therapy, like siRNA therapy, is limited by inefficient delivery to target cells and off-target effects and further work is under way to improve the usefulness of these molecules as therapeutic agents.

How do we study RNA?

There are a number of different laboratory techniques that can be used to study RNA molecules. One of the first techniques to be developed was Northern blotting. This method is similar to Southern blotting, but instead of identifying DNA, specific RNA molecules from a mixture are detected. First the RNA is extracted from a particular cell or tissue type. The RNAs within the sample are separated by size using electrophoresis and then transferred onto a membrane to which the full set of extracted RNAs becomes attached. A labelled probe, complementary in sequence to that of the RNA sequence being investigated, is then added to the membrane. On exposure of the blot to specific conditions, the RNA molecule to which the probe has hybridized is identified.

The Northern blotting technique has now largely been replaced by more rapid and sensitive techniques such as RT-PCR. In RT-PCR, mRNA is extracted from cells or tissues, converted into cDNA, and subsequently amplified. To quantify the amount of gene (mRNA) being expressed an extension to the technique called real-time or quantitative RT-PCR (qRT-PCR) can be used. This is useful when comparing gene expression between different cell types or in cells under different physiological or experimental conditions.

Microarrays

Both Northern blotting and RT-PCR enable the expression of one or a few genes to be measured simultaneously. In contrast, the technique of microarrays allows gene expression to be measured across the full genome of an organism in a single step. This massive scale genome analysis technique is very useful when comparing gene expression profiles between two samples. For instance between healthy versus diseased tissue, cells administered with a drug versus no drug treatment, or cells infected by pathogens versus those that are unaffected. This can identify gene subsets that are under- or overexpressed in one sample relative to the second sample to which it is being compared. In this method, mRNA is extracted from both samples and converted to cDNA using reverse transcriptase. To differentiate the samples, cDNA from one sample is labelled with a red fluorescent marker and the second with a green fluorescent marker. The two samples are combined and then added to a DNA microarray, or DNA chip. These are DNA sequences representing different genes fixed to a miniature support such as a glass slide. When the combined samples are added, the cDNA binds to complementary sequences on the array. The microarray is scanned for fluorescence and captured as images using a fluorescent microscope. The intensity of fluorescence for each DNA sequence represents the number of labelled cDNA molecules bound to that sequence and thus the amount of mRNA present in the original sample. By analysing this data computationally, relative differences in gene expression between the two samples can be quantified.

RNA sequencing

RNA sequencing, also termed RNA-seq, is fast becoming the technique to identify RNA molecules and measure gene expression profiles. In this method, total RNA is extracted, fragmented, converted into cDNA, and then sequenced using next-generation sequencing approaches. Instead of extracting total RNA, which includes all the different RNA molecules found in a cell, it is also possible to extract specific populations such as miRNA or lncRNAs. These are isolated based on size differences and converted to cDNA prior to sequencing. RNA-seq has been used in a number of large-scale projects. It was used to identify and catalogue functional elements within the genome as part of the ENCODE project and has transformed our understanding of the genome and how it works. It has also been used as part of the Cancer Genome Atlas project to measure differences in gene expression profiles between cancer and non-cancer cells. Data from this study is paving the way for new ways of diagnosing and treating cancers.

Chapter 4
Proteins

Biological functions require protein and the protein makeup of a cell determines its behaviour and identity. Not surprisingly therefore proteins are the most abundant molecules in the body except for water. Protein is derived from the Greek word *proteios* meaning first or primary as it was considered to be the primary form of nutrition made by herbivores. The approximately 20,000 protein-coding genes in the human genome can, by alternative splicing, multiple translation starts, and post-translational modifications, produce over 1,000,000 different proteins, collectively called 'the proteome'. It is the size of the proteome and not the genome that defines the complexity of an organism. Of all the organs in the body, testicles have been found to contain the highest number of unique proteins.

Proteins make up half the dry weight of a cell whereas DNA and RNA make up only 3 per cent and 20 per cent respectively. The link between nucleic acid and protein received scant attention until 1941, when George Beadle and Edward Tatum showed unequivocally that genes direct the manufacture of proteins, which in turn control metabolism. These pioneers created mutant bread moulds that required the addition of the amino acid arginine in order to grow. Each mutant had lost one gene and each mutant was shown to lack one enzyme involved in the synthesis of arginine. This led to the 'one gene one enzyme' hypothesis.

Enzymes are proteins that catalyse or alter the rate of chemical reactions, and by showing that genes controlled the production of enzymes Beadle and Tatum revealed for the first time how genes were playing a major role in molecular biology.

Enzymes can speed up reactions that would otherwise take a very long time to complete at body temperature but they also slow some reactions down. Proteins play a number of other critical roles. They are involved in maintaining cell shape and providing structural support to connective tissues like cartilage and bone. Specialized proteins such as actin and myosin are required to provide contraction for skeletal and cardiac muscular movement. Other proteins act as 'messengers' relaying signals to regulate and coordinate various cell processes, e.g. the hormone insulin. Yet another class of protein is the antibodies, produced in response to foreign agents such as bacteria, fungi, and viruses.

Composition of proteins

Proteins are composed of amino acids. Amino acids are organic compounds with two essential features, an amino group ($-NH_2$) at one end (N terminus) and a carboxyl group ($-COOH$) at the other (C terminus). In addition, amino acids carry various side chains that give them their individual functions. The twenty-two amino acids found in proteins are called proteinogenic (twenty universal and two newly discovered amino acids) but other amino acids exist that are non-protein functioning. An example is gamma amino butyric acid (GABA), a neurotransmitter that is synthesized from glutamate in the brain. Nine of the proteinogenic amino acids are termed 'essential' for humans since they cannot be synthesized by the body and must therefore be ingested in the diet.

Attachment of the amino group from one amino acid to the carboxyl of another forms a chemical bond linking the two amino

12. A peptide bond is formed between two amino acids by the removal of a water molecule.

acids together. The formation of this peptide bond is achieved by the removal of a water molecule, therefore each individual unit in a peptide or protein is known as an amino acid residue (see Figure 12). Chains of less than 50–70 amino acid residues are known as peptides or polypeptides and >50–70 as proteins, although many proteins are composed of more than one polypeptide chain. In 1949 Frederick Sanger sequenced the first protein, insulin, for which he was awarded the Nobel Prize in 1958. Importantly this demonstrated conclusively that proteins consisted of linear chains rather than branched chain structures such as in starches.

Proteins are macromolecules consisting of one or more strings of amino acids folded into highly specific 3D-structures. Each amino acid has a different size and carries a different side group. It is the nature of the different side groups that facilitates the correct folding of a polypeptide chain into a functional tertiary protein structure. Hydrogen bonding, the weak attractive force between a hydrogen atom in a molecule and a nearby negatively charged atom, was first proposed by Linus Pauling as a mechanism for promoting protein folding into a tertiary structure. Folding is also mediated by hydrophobic interactions as first shown by Walter Kauzmann and Kaj Linderstrom-Lang. Hydrophobic or 'water-hating' side chains on amino acids cluster together, buried in the centre of a molecule, and thus avoid contact with water. Correct protein folding also involves special proteins called molecular chaperones that catalyse the bending and folding of a protein. Proteins are thus composed of polypeptide chains, acting either singly or as multiple peptide subunits. Many proteins also have co-factors attached such as metal ions or organic groups such as the haeme group in haemoglobin. Organic co-factors are often derived from vitamins such as vitamin B6, niacin, or folic acid.

When a protein function is no longer required or when they become fatigued, proteins are marked for destruction, broken down, and their components recycled. This protein turnover is carefully managed so that damaged proteins or those with functions that are only temporarily required are not retained to interfere with the functioning of the cell. A protein's lifespan is measured in terms of its half-life, which can last as little as a few seconds or as long as years. Inefficient or incomplete protein turnover can lead to abnormal accumulation of substances in cells that cause, for example, Alzheimer's disease.

How do we study proteins?

What information do we require in order to understand how proteins work and what happens when this function is impaired in disease?

First, it is important to be able to isolate the protein of interest from the 'cellular soup' of macromolecules. Vital for purification is a means of identifying the protein of interest throughout the isolation process. This can be achieved on the basis of the overall charge of a protein, which is due to the type of amino acids at its surface, or alternatively by its size, which depends on the number of amino acids the protein contains, or a combination of the two.

Isoelectric focusing, 1D and 2D electrophoresis

Amino acids vary in the charge they carry at physiological pH (approximately pH7) and in turn this affects the surface charge on a protein. In the laboratory we can change the pH of the environment so that the negatively charged amino acids balance the net effect of the positively charged ones. The pH at which there is no net charge on a protein is known as the isoelectric point (pI) and is an identifying feature of a protein.

Proteins are separated on the basis of their surface charge by being applied to gels in which a pH gradient has been immobilized. The gel material used is usually acrylamide, which has a uniform and easily controlled pore size. Proteins move through the gel under the influence of an electric current until they come to rest at the part of the gel in which the pH matches their pI. At this pH the protein has no net charge and therefore migration ceases. The result is a gel with banded proteins, each positioned according to their pI.

Proteins can also be separated according to their size by molecular sieving or electrophoresis. To do this proteins are coated in detergent so they each have an identical surface charge. This enables them to be separated, based purely on their size. To speed up the separation process proteins are pulled through the sieving matrix by an electric current. The faster moving smaller proteins move ahead of the larger molecules that move more slowly in the gel. These two separation techniques based on the different parameters of size and charge can be combined in a two-dimensional method known as 2D gel electrophoresis. First proteins are separated based on charge, followed by a further separation due to protein size. The proteins can be visualized in

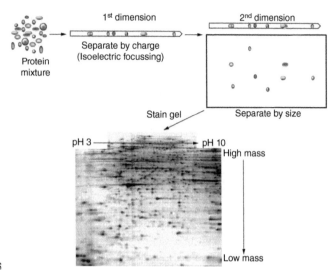

13. Two-dimensional gel electrophoresis can separate proteins on the basis of their charge and then their size (mass). Proteins are then stained with silver (in this case) to reveal a unique pattern of spots, as shown in the lower panel.

the gel by special stains (see Figure 13). The bands or spots can be recorded or excised for further analyses such as by mass spectrometry (MS) if required.

Mass spectrometry—weighing molecules

Proteins are characterized by means of their mass, or how big they are, in conjunction with their overall or net electric charge. This is their mass to charge ratio, which can be measured by the technique of mass spectrometry (MS). The basic principle of MS is that the amount a moving object can be deflected by a given force depends on its mass. An analogy would be if you hit a plastic table-tennis ball it would deflect from its path by a greater distance than a tennis ball, given the same strength of hit. In place of the bat or racquet, magnetic fields can be used to deflect the electrically charged particles known as molecular ions. The technique can be used to identify a number of proteins within a mixture to all of the proteins found in a cell using only small amounts of starting material. It can also be employed to analyse previously isolated proteins. MS has many uses: in medicine for drug testing or neonatal screening using biological fluids or tissues, to identify environmental contamination such as pollutants in rivers, and it is used by the pharmaceutical industries when testing the properties of new drugs.

Studying the proteome

Marc Wilkins first used the term 'proteome' in 1994 to refer to all the proteins expressed by a cell, tissue, or organism under defined conditions. For simple organisms, such as viruses, all the proteins coded by their genome can be deduced from its sequence and these comprise the viral proteome. However for higher organisms the complete proteome is far larger than the genome due to alternative splicing of genes, different start and stop sites for translation, and post-translational modifications. For these organisms not all the proteins coded by the genome are found in any one tissue at any one time and therefore a partial proteome is usually studied. What are of interest are those proteins that are expressed in specific cell types under defined conditions.

Proteomics, or the study of the proteome, is usually carried out by 2D-electrophoresis since this can separate 2,000 peptides. Peptides can be identified using their mass and isoelectric point as coordinates by comparison with standards in databases. More recently unknown peptides have been identified using mass spectrometry and a technique called peptide mass fingerprinting, revolutionizing the study of proteomes. An advantage of this method is that only the masses and not the sequences of the peptides need to be determined. A draft map of the human proteome has recently been published using this strategy (1). The authors amalgamated the proteomic data from thirty different normal human tissues enabling them to identify the proteins encoded by 17,294 genes; an estimated 84 per cent of the protein-coding genes in the human genome. The significance of the normal proteome is that it will provide the basis for research into the ways proteins change with disease.

Studying protein structure

The functions of most proteins rely on 3D structure and the potential for interaction with other macromolecules or co-factors that it provides. To study this we need some means of magnification that allows us to visualize protein structure at the atomic level. This requires a form of electromagnetic radiation with wavelengths short enough to distinguish atoms; visual light wavelength is too long. X-rays have the properties of electromagnetic radiation with wavelengths down to 0.1 nm, perfect for looking at atomic structures. The birth of this technology occurred approximately 100 years ago due to the German scientist Max von Laue who discovered X-ray diffraction of crystals. When X-rays are passed through the regular array of atoms in a crystal they scatter and the resultant different waves then interfere with each other either by addition or by cancellation. It is a bit like the waves on a pond when a stone is dropped in. The X-ray diffraction pattern produced can be used to back-calculate the arrangement of atoms in the original molecules as we have seen for DNA.

Purified proteins can form crystals and as such have a regular array of atoms that can scatter X-rays to form a diffraction pattern. Atoms scatter the waves of X-rays mainly through their electrons, thus forming secondary or reflected waves. The pattern of X-rays diffracted by the atoms in the protein can be captured on a photographic plate or an image sensor such as a charge coupled device placed behind the crystal. The pattern and relative intensity of the spots on the diffraction image are then used to calculate the arrangement of atoms in the original protein. Complex data processing is required to convert the series of 2D diffraction or scatter patterns into a 3D image of the protein. The first protein structures to be solved were myoglobin and haemoglobin in 1958. The continued success and significance of this technique for molecular biology is witnessed by the fact that almost 100,000 structures of biological molecules have been determined this way, of which most are proteins. X-ray crystallography is also used in industry, for example in pharmaceuticals to monitor drug protein interactions.

Once a key protein in a disease or infection has been identified as a potential target for a drug we need to see how it works on an atomic scale. X-ray crystallography enables 'visualization' of the 3D structure of a protein thus providing key information on the arrangement of amino acids in its active site, the region where molecules bind and then undergo a chemical reaction. Having identified the amino acids in a target site, drugs can be designed that will bind and block activity of the crucial protein. Molecular biology in combination with X-ray crystallography led to discovery of new treatment for hepatitis C (HCV), a virus that affects over 170 million people worldwide. Previous standard therapy for chronic HCV involved injections of the antiviral agent interferon alpha that had considerable side effects. The X-ray structure of a key enzyme that mediated infection called serine protease was carried out. Having identified the key amino acids in the active site of this enzyme, small peptides were designed that would fit into this site and inhibit the viral protease. This serine protease is crucial for cleaving a protein coded by the RNA genome of HCV and its inhibition therefore prevents replication of the virus. The serine protease has a dual role as it also destroys part of the host's antiviral defences. The drug that was made possible by analysis of protein structures is now marketed under the name Telaprevir^{TM}.

The world Protein Data Bank (PDB) archives almost 100,000 protein structures, collected since 1971. Multi-protein complexes are now being attempted by crystallizing them piecemeal and then joining the individual parts together rather like a jigsaw puzzle. Specialized lasers are replacing the X-ray beam. Important structures solved to date include the HIV trimeric hook that the virus uses to bind to and infect its preferred cell for replication. This complex protein structure in the envelope of the virus looks like the three legs of Mann, the famous symbol for the Isle of Man.

Many proteins do not form large and stable crystals and are therefore unsuitable for this technology. Despite the success of X-ray crystallography, solution nuclear magnetic resonance spectroscopy (NMR) is now being used as an alternative and complementary approach to characterizing molecular structures. NMR exploits the magnetic properties of atomic nuclei, which absorb and re-emit electromagnetic radiation according to the local molecular environment that affects the nuclei of atoms. The technique does not generate an image directly but depends on calculation to generate 3D models. It could be imagined as a kind of MRI scan for molecules that enables scientists to calculate the structure of a molecule. It is used to determine not only molecular structures but also their interactions with other molecules, even if weak or transient. One of its major benefits is that structures can be determined with the molecules in solution, which has greater physiological relevance.

Protein identification by immunological means

An antibody is known as immunoglobulin or IgG. Immunoglobulins have highly specific molecular recognition properties and bind in a 'lock and key' fashion to precise sites or epitopes on another protein, called an antigen. This high specificity has been appreciated as a powerful tool that is fundamental to many identification techniques in molecular biology.

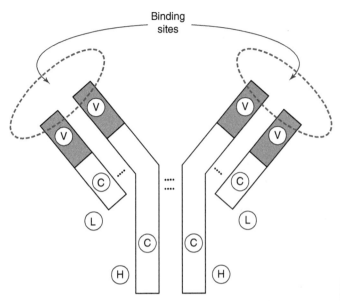

14. An immunoglobulin molecule comprising two heavy (H) and two (L) light polypeptide chains forming a 'Y'-shaped structure with a constant end (C) and two variable antigen-binding sites (V).

Each immunoglobulin molecule is composed of four polypeptides, two large ones called heavy chains and two smaller ones called light chains. These are covalently bound in a 'Y'-shaped configuration. The tips of the 'Y' are hypervariable, forming binding sites for a wide range of antigens. The 'stalk' of the 'Y' on the other hand is constant within a species (see Figure 14). Immunoglobulin IgG is a member of a superfamily of immunoglobulin proteins involved in a range of functions in the immune system.

Western blot

Western blot is an immunological technique that identifies proteins after gel-separation. Harry Towbin of the Friedrich Miescher Institute Switzerland introduced this technique in the

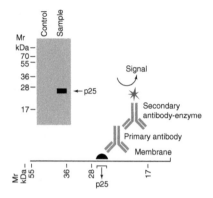

15. Western blot technique uses a combination of separation by size, blotting on to a membrane and probing using specific antibodies to identify a protein in a mixture.

1970s and with tens of thousands of primary antibodies now commercially available it has become one of the mainstays of molecular biology. It was named 'Western' in line with the 'Southern' DNA blots of Edwin Southern and the RNA blots that were nicknamed 'Northern'. There is no Eastern blot.

For a Western or protein immunoblot, separated proteins are transferred from the acrylamide gel by moving or blotting them onto a membrane in order to make them accessible to the antibodies employed in specific recognition. Since the membrane has a high affinity for proteins any excess binding capacity must then be blocked before probing with the specific or primary antibodies which themselves are protein. The primary antibody or IgG binds to any protein on the blot containing its specific epitope. The bound primary is then detected by means of a secondary antibody that recognizes and binds to the IgG constant region of the primary. The secondary antibody comes tagged with an enzyme such as peroxidase or alkaline phosphatase that then acts on a chemical to convert it into a coloured or chemiluminescent end product thus enabling the protein of interest to be identified and quantified (see Figure 15).

56

Immunohistochemistry (IHC)

Studying component proteins after a tissue has been macerated has been likened to determining the menu after the contents of your plate has been made into a soup. Much important information will have been lost such as where in the cell the protein is located or even in which cell types. Diagnoses of diseases are often made using thin sections of tissue and identifying proteins associated with specific cell types, functions, or disease states. Study of protein expression in individual cells of a tissue is also fundamental to biomedical research. In this technique the proteins in the tissue or organ are fixed or cross-linked to preserve their structure and then the material is embedded in a matrix such as paraffin wax. This allows sections of the tissue to be cut with ease. Thin sections of the tissue are then supported on glass microscope slides and stained for the presence of specific proteins. The immunological properties of the protein are usually retained in fixed tissue and can be used with specific antibodies to identify a protein in a similar way to Western blotting. In IHC, the secondary antibody can be conjugated to a fluorophore that provides amplification of the signal when activated to fluoresce at the microscope. Alternatively the secondary antibody can be conjugated to an enzyme that produces a coloured product at the site of the protein of interest which can be visualized by a light microscope. A counterstain is often applied to highlight features such as the nucleus and cell membrane before examination and storage.

Gene duplication and protein paralogues

Errors in DNA replication have over millions of years produced duplicate copies of certain ancestral genes. These can then evolve separately to produce differently functioning proteins or paralogues. Gene duplication is another important mechanism in the evolution of complex organisms leading to formation of protein paralogues or isoforms. A good example of this is the enzyme enolase that functions in the energy-producing process of glycolysis. This enzyme has three isoforms in vertebrates, alpha, beta, and gamma, that result from two gene duplication events. Enolase is highly expressed in glycolytic tissues such as brain, muscle, and liver that have demanding energy requirements. It catalyses the penultimate step in glycolysis by conversion of 2 phosphoglycerate (2PG) to phosphoenolpyruvate (PEP). The alpha form is widely expressed and is the only form in liver. Beta enolase is specific for muscle and gamma enolase for neuronal and neuroendocrine cells. In this way different tissues can express different forms of the same protein. The enolase paralogues arose from a single gene duplication event early in vertebrate evolution followed by a second duplication that gave rise to separate beta and gamma isoforms. Beta and gamma proteins are the most closely related of the three isoforms.

Proteins with multiple functions—'moonlighting'

The human is considered the most complex organism in nature and this is reflected in the relative size of the human proteome, which is three times as complex as that of a fly. The number of proteins in higher organisms far exceeds the number of known coding genes. The fact that many proteins carry out multiple functions but in a regulated manner is one way a complex proteome arises without increasing the number of genes. Proteins that performed a single role in the ancestral organism have acquired extra and often disparate functions through evolution.

Piatigorsky and Wistow in 1987 made the astonishing discovery that the clear proteins in the vertebrate lens of the eye, known as crystallins, were in fact the same proteins as some of the glycolytic enzymes involved in breaking down glucose to produce energy. Epsilon-crystallin in the eye of a duck is the enzyme lactate dehydrogenase, and in the turtle lens tau-crystallin was shown to be the glycolytic enzyme alpha-enolase. These enzymes accumulate in the lens to very high concentrations where a metabolic role seems unlikely. Instead they probably have only a structural function. The active site of an enzyme employed in catalysis is only a small part of the protein, leaving spare capacity for acquiring a second function.

Enzymes that have roles in addition to catalysis are among the most common examples of this gene sharing. The glycolytic pathway is involved in the breakdown of sugars such as glucose to release energy. Many of the highly conserved and ancient enzymes from this pathway have developed secondary or 'moonlighting' functions. Proteins often change their location in the cell in order to perform a 'second job'. Thus in the cytoplasm the enzyme enolase performs its catalytic function for the glycolytic pathway, in the lens it is structural, but when located at the cellular membrane it has been shown to be involved in pro-inflammatory responses.

The limited size of the genome may not be the only evolutionary pressure for proteins to moonlight. Combining two functions in one protein can have the advantage of coordinating multiple activities in a cell, enabling it to respond quickly to changes in the environment without the need for lengthy transcription and translational processes. Molecular biologists interested in cell adhesion and the organization of the cytoskeleton first described this when they were working on a protein called beta-catenin. Beta-catenin is widely expressed at the cell surface and is involved in maintaining cell shape and epithelial integrity. Imagine their surprise when they found that it was the very same protein that was being studied for its transcriptional activity within the nucleus. The dual role for this protein makes biological sense as it coordinates cell surface signalling with transcriptional responses. Instead of just destroying the beta-catenin that has been displaced from the cell surface it is used as a transcriptional activator. This is an ingenious example of a moonlighting protein that links a structural change at the cell surface with a signalling pathway.

A loss of cell–cell adhesion releases beta-catenin where it is transported into the nucleus to carry out its second role as a transcription factor activating a battery of genes necessary for proliferation.

Protein isoforms produced from a single gene

Protein isoforms are different forms of essentially the same protein that can differ in their location either within a cell or between different cell types. RNA from a single gene can be differently processed to form several highly related gene products or protein isoforms that have diverse functions. Alternative splicing of the pre-mRNA is one means by which different proteins can be produced from the same gene. Protein isoforms are also created by the use of alternative start and stop signals during translation to give rise to families of proteins with similar functions but regulatory differences. A good example of this is the gene *TP53*. *TP53* codes for a family of proteins; the full-length protein FLp53 and eleven shorter isoforms created using alternative start sites and splicing of exons to produce the mRNAs (2). The biological significance of all of the twelve isoforms has not yet been determined but it is certain that they play a role in translating stress signals into homeostatic survival behaviour.

Post-translational modifications

Post-translational modification (PTMs) as the term suggests is another process that can modify the role of a protein by addition of chemical groups to amino acids in the peptide chain after translation. Addition of phosphate groups (phosphorylation), for example, is a common mechanism for activating or deactivating an enzyme. Other common PTMs include addition of acetyl groups (acetylation), glucose (glycosylation), or methyl groups (methylation). Small proteins can also be added such as ubiquitin (ubiquitination) that can change the function of a protein. Some additions are reversible, facilitating the switching between active and inactive states, and others are irreversible such as marking a protein for destruction by ubiquitin.

Diseases caused by malfunction of these modifications highlight the importance of PTMs. Alzheimer's disease, for example, is characterized by the failure to 'recycle' structural proteins in the brain. The resultant misfolded protein builds up in nerve cells, impairing their function and eventually leading to cell death and neuron loss. Another example occurs in diabetes. High blood glucose levels lead to unwanted glycosylation of proteins. At the high glucose concentrations associated with diabetes, an unwanted irreversible chemical reaction binds the glucose to amino acid residues such as lysines exposed on the protein surface. The glycosylated proteins then behave badly, cross-linking themselves to the extracellular matrix. This is particularly dangerous in the kidney where it decreases function and can lead to renal failure.

Prions

Prions are proteins that require a special mention for when misfolded they can cause the devastating transmissible spongiform 'infections' such as 'Mad Cow' disease.

Stanley Prusiner derived the term 'prion' in 1982 from the words '*Pro*tein' and 'Infect*ion*'. Prions are responsible for these untreatable and fatal infectious diseases of nervous tissue such as Creutzfeldt-Jakob disease and its variant, the so-called 'Mad Cow' disease. A prion, or prion protein, is an incorrectly folded version of a natural occurring protein. When the normal version of the protein comes into contact with the prion it causes the normal protein to become misfolded, thus triggering a chain reaction. The disease relies on the presence of the normal protein for its propagation. The pathological form of the prion protein (PrP^{Sc}) is misfolded into an aggregate that cannot be removed by protein turnover machinery and its accumulation causes cell death.

This gives rise to the characteristic spongy appearance of the infected neuronal tissue. In humans the normal protein (PrP^C) of only 209 amino acids is membrane bound where it is believed to function in cell adhesion and communication. Prions replicate by attaching normal PrP^C to the ends of the aggregated PrP^{Sc} fibrils thus converting them to the pathological conformation. Since prion propagation involves changes in protein conformation only, leaving the gene sequence unchanged, prions are not technically an exception to the central dogma of DNA to RNA to protein.

Chapter 5
Molecular interactions

Every nucleated diploid cell in the body, with the exception of
B and T cells of the immune system, has the same genome as its
originating single fertilized egg. During development, this single
cell differentiates into a complex multicellular organism comprising
various cells and tissues each carrying out specialized functions.
Although each cell contains a genome of data it needs to select the
relevant information from this genetic blueprint to fulfil its own
specific function. Proteins must be produced in the right place and
at the right time. This requires regulation of gene expression in
conjunction with a myriad of bio-molecular interactions to
coordinate this. These interactions require the synchronized activity
between proteins, RNAs, and DNA and molecular biology lies at
the core of our understanding of these control processes.

Regulation at the chromatin level

Only subsets of the genes are active in each cell type and at each
stage of development. The pattern of gene expression determines
the behaviour and identity of the cell, for example whether it is a
white blood cell, a sperm, or a liver cell. Gene expression is
regulated at many stages, of which one is at the level of
transcription. Transcriptional regulation occurs primarily at the
pre-initiation stage. This is achieved initially at the level of the
chromatin by controlling access to the DNA coding region for the

transcriptional machinery. Chromatin is made up of DNA and histone proteins that not only stabilize the DNA but also play a role in transcriptional control. Chromatin compaction by the histones leads to gene silencing, and for a gene to be transcribed the chromosome structure around it needs to be unwound to allow access for transcription factors (TF) and ancillary molecules. Transcription factors are proteins that bind to specific sequences on DNA to promote or inhibit transcription of nearby genes. A key mechanism for silencing genes, especially during embryonic development, is the remodelling of chromatin by special protein complexes. These are required for histone modification that leads to long-term chromatin silencing in differentiating cells.

Chromatin remodelling to regulate gene expression also occurs on a temporary basis by addition or removal of chemical groups such as methyl and acetyl to histone proteins as a dynamic mechanism for keeping genes active or silent. This is called the histone code. Histone modifications such as acetylation by specific enzymes loosen chromatin compaction allowing access of TFs to transcriptional start sites and subsequently induce gene transcription. In contrast deacetylation, again by specific enzymes, increases chromatin compaction and causes recruitment of proteins that inhibit transcription.

Regulation at transcriptional level

The ease by which RNA polymerase and TFs can access a gene is only one element in the regulation of gene expression. TFs can act as activators or repressors according to the nature of the co-factors they recruit to the DNA. Many different mechanisms exist to control the activity of a TF, one of which is intracellular localization. TFs must act in the nucleus, so they can be maintained in an inactive form by retaining them in the cytoplasm. Rapid activation is achieved by their release from the cytoplasmic inhibitory complex enabling them to enter the nucleus and transcribe their specific repertoire of genes. This is a means of rapid activation of urgently required genes. An example is nuclear factor kappa-light-chain-enhancer of activated B cells (NFkB) that is held in an inactive complex in the cytoplasm until required for a stress response such as elicited by infection. The activation of NFkB involves proteolytic destruction of its inhibitor, freeing it for entry into the nucleus where it can activate transcription of relevant genes. In the case of NFkB these are pro-inflammatory genes. Humans express at least one TF for every ten genes. TFs govern the way genes are transcribed and how RNA polymerase is recruited. TFs combine with other functional and regulatory proteins to express the specific pool of proteins required by the cell.

In the laboratory, techniques such as chromatin immunoprecipitation (CHiP) are used to determine the DNA sequences that TFs bind to. In CHiP, DNA and its associated proteins (chromatin) are first fixed to cross-link the DNA and attached proteins together. With DNA and proteins stably bound the chromatin is then broken into smaller fragments, either mechanically or using enzyme digestion. These smaller fragments can then be selectively immunoprecipitated using antibodies specific to the DNA binding protein of interest. The precipitated chromatin fragments are then purified and the DNA sequenced. In this way the genomic sequences that a protein of interest binds to can be determined.

Regulation by enhancer sequences

Enhancer sequences are non-coding regions of DNA, between 200 base pairs and 1,000 base pairs in length that occur in the region of a promoter. Enhancer or repressor molecules can bind to enhancer sequences and influence the ability of a TF to activate a promoter. A promoter may have as many as four or five enhancer sequences located either near the promoter or even at a distance since 'DNA looping' can bring them into close proximity. Even when the RNA polymerase is bound to a promoter it still requires another set of factors to allow it to move off the promoter complex and begin successfully transcribing RNA. There is another level of control in the determination of when the RNA transcript is terminated.

Regulation by epigenetics

Twenty thousand protein-coding genes make up the human genome but for any given cell only about half of these are expressed. Some genes are required for 'housekeeping' by almost every cell but others are specific for the cell lineage, such as haemoglobin in red blood cells. Many genes get switched off during differentiation and a major mechanism for this is epigenetics. The term epigenetics comes from the Greek *epi-* meaning 'upon', to signify a layer of control that is laid upon or acts above the DNA code. This is the mechanism whereby genes are 'marked' for expression or not as part of a selection process. Epigenetics programmes the cell so that it can select the information in the genome required for its functional specialty. The basic sequence of the genetic code remains intact and is passed on to daughter cells during cell division along with any specialized epigenetic changes. A consensus definition of an epigenetic trait given at the 2008 meeting in Cold Spring Harbor, USA, is 'a stably heritable phenotype resulting from changes in a chromosome without alterations in the DNA sequence'. Epigenetics involves the chemical alteration of DNA by methyl or other small molecular groups to affect the accessibility of a gene by the transcription machinery (see Figure 16).

The genomic sequence is conserved in all cell types but the epigenomic landscape can vary considerably, contributing to the distinct gene expression required for biological functions.

The case of the Agouti mouse clearly demonstrates the fact that changes in phenotype or appearance can be caused by means other than DNA genetics. Identically genetic mice carrying a gene for obesity and yellow fur (Agouti) can either be black if their

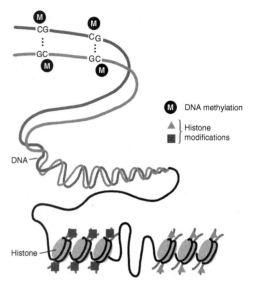

16. Epigenetic variations consist of DNA methylation and histone modifications.

mother is fed a diet inclusive of the B vitamin folate or yellow if the maternal diet lacks folate. The vitamin folate is a key source of methyl groups used to methylate DNA. Deprivation of folate and hence methylation prevents the agouti gene in the embryos from deactivation and the offspring have yellow fur and become obese, diabetic, and with a high risk of cancer even if fed on a normal diet. In the presence of folate, and therefore methylation, embryos carrying the agouti gene can switch it off by modifying the DNA.

Maintenance of epigenetic marks during cell division

DNA modification by methylation results from the conversion of cytosine to 5-methylcytosine by enzymes, aptly called DNA methyltransferases (DNMTs). This is an epigenetic modification, as it doesn't alter the genetic code embedded in the original DNA sequence. The methylated cytosine residues are usually immediately adjacent to a guanine nucleotide (CG), resulting in two methylated cytosine residues sitting diagonally opposite on opposing DNA strands (see Figure 16). During cell division the semiconservative nature of DNA replication initially gives rise to hemi-methylated DNA where the parent strand retains the methylation marks and the newly synthesized strand is unmethylated. Immediately after DNA synthesis a DNA methyltransferase, specific for maintaining methylation signals, recognizes these hemi-methylated CG sites and methylates the newly incorporated cytosine. This maintains the identical methylation pattern of the parental DNA. In this way the epigenetic marks and the pattern of gene regulation that existed in the parental cell is retained in the daughter cells and throughout subsequent multiple cell divisions.

Epigenetic changes determine whether particular genetic information will be read, and then when and how. Epigenetic inheritance is a means of ensuring the transmission of epigenetic marks, from parental to daughter cell and potentially from generation to generation.

Heritability of epigenetic marks

Epigenetic signatures are not only passed on to somatic daughter cells but they can also be transferred through the germline to the offspring. This has surprised scientists as most DNA methylation is stripped in the embryo and the mechanisms for heritability of an epigenotype are not as yet understood. However in the last decade scientists have begun to report examples of exposure to environmental factors in parents being passed on to their offspring by changes in DNA methylation.

In 2005 it was reported that British men who started smoking before puberty had an increased risk of fathering boys of above average weight. Similarly Swedish men who had experienced famine in childhood conferred a reduced risk of heart disease or diabetes on their grandsons. The question arises as to whether environmental factors such as neglectful parenting, diet, or airborne pollutants can affect subsequent generations through the generation of heritable but epigenetic changes to the germline DNA. At first the evidence appeared circumstantial but more recent studies have provided direct proof of epigenetic changes involving gene methylation being inherited. Rodent models have provided mechanistic evidence. Males fed high-fat diets had female offspring with altered methylation of the DNA in the pancreas. Low-protein diets in male mice led to fathering of offspring with altered expression of genes in cholesterol metabolism in their livers. Pre-diabetic mice had altered methylation of sperm DNA that caused increased risk of the next two generations of offspring developing diabetes.

Further evidence comes from a startling but controversial study published in 2013 by Brian Dias and co-workers showing mice that associated fear with the chemical acetophenone, passed on this information to their offspring (1). Acetophenone binds to a nasal receptor encoded by the gene with the name *Olfr151*. Exposure to the frightening smell resulted in epigenetic changes to this gene in the DNA of their sperm. This led to increased levels of the specific nasal receptor and enhanced sensitivity to the smell. In this way a fear of a perceived dangerous odour is passed on to future generations. It is possible that a similar phenomenon could be responsible for transmitting some fears between the generations in humans. Although the mechanism behind inheritance of epigenetic signals is uncertain the importance of epigenetics in development is highlighted by the fact that low dietary folate, a nutrient essential for methylation, has been linked to higher risk of birth defects in the offspring.

Short RNA molecules that bind to DNA and affect gene expression have also been implicated in passing on environmental effects to offspring. Twenty-eight such microRNAs are differentially expressed in sperm of men who smoke and those who do not.

An Australian study led by Michelle Lane found that obese male mice could pass on insulin resistance to the next two generations through abnormal microRNA expression in their sperm.

Gene–environment interactions

The role of enhancers and epigenetics in mediating changes in gene expression and hence vulnerability to disorders is widely accepted but how these interact with the environment is now becoming of great importance and also controversy. The first gene to be linked with social behaviour is a particular variation of the X chromosome gene *MAOA* that produces the enzyme monoamine oxidase A. This enzyme degrades the neurotransmitters dopamine, norepinephrine, and serotonin in the brain affecting their levels and therefore our moods. Inhibitors of MAOA activity are common drugs used in treatment of clinical depression and anxiety. Mutations in *MAOA* cause the Brunner Syndrome, characterized by violent and impulsive behaviour. The lack of activity of the mutant enzyme gives rise to an excess of monoamine neurotransmitters, causing the behaviour seen in this syndrome. Levels of expression of the wild type gene can also cause behavioural differences. The *MAOA* gene has two types of enhancer sequence upstream of the promoter. The shorter form is associated with low enzyme expression and has been named the 'Warrior gene' as it enhances the qualities associated with successful fighters. However this low expression variant gene in combination with maltreatment during childhood may result in an increased risk of antisocial and aggressive behaviour. Studies have linked methylation of the *MAOA* gene and low expression with nicotine and alcohol dependence in women but not in men. Interestingly epigenetic methylation of this gene is very low in men.

Epigenetics and X-inactivation

Females have two copies of the X chromosome and males only one, so to avoid overactivity of X-linked genes in females one copy is inactivated. Epigenetic signalling by high levels of DNA methylation is involved in silencing one copy of the X chromosome. The inactive X chromosome is packaged in compact chromatin or heterochromatin that keeps most of the DNA transcriptionally silent. The long non-coding RNA Xist (X-inactive specific transcript), expressed only from the inactive X chromosome, coats this chromosome, thus silencing its other genes. Faulty X-inactivation or skewed X-inactivation can occur, and this has been associated with various diseases including autoimmunity, autism, and cancer. X-inactivation is reversed in female gametes so all egg cells have both copies of the X chromosome active. X-inactivation is random in placental mammals but is always applied to the paternally derived X chromosome in marsupials. Interestingly embryonic cells destined to become part of the placenta all have the paternally derived X chromosome inactivated.

An example of X-inactivation is commonly seen in tortoiseshell cats. A tortoiseshell cat has black and ginger or orange patches in a pattern reminiscent of the shell of a tortoise. This coat colouration only occurs in females, since the genes for coat colour lie on the X chromosome. The random inactivation of either the black coding 'X' or the ginger coding 'X' chromosome results in the characteristic mosaic-effect of the tortoiseshell coat colour. The process of random X-inactivation, or lionization, takes place early in embryonic development known as the blastula stage. All the skin cells that descend from a single blastula cell or blastomere will express the same X-inactivation. Some of these blastomeres develop into pigment-producing cells or melanocytes that migrate to the skin where they intermingle to give the characteristic brindle appearance.

Epigenetics can therefore act on gene expression without affecting the stability of the genetic code by modifying the DNA, the histones in chromatin, or a whole chromosome.

Gene expression and beyond

The expression of individual genes must also respond to internal and environmental stresses in order to provide the necessary proteins for the functioning of signalling and metabolic pathways. These genes are under a tightly regulated combination of controls that can involve transcriptional activators and repressors, microRNAs, histone remodelling, and epigenetic modifications such as methylation of DNA.

Molecular interactions that maintain cell number

Controlling gene expression is not only important for the development of the embryo but also for its maintenance or homeostasis throughout life. Take skin, for example; in spite of around 40,000 cells dying every hour we maintain an intact and smooth covering. The interface between cell death and cell renewal is key to keeping the correct number of cells in the body and this requires timely regulation of gene expression in addition to a myriad of molecular interactions. Loss of gene regulation and consequent breakdown of the link between cell death and cell cycle processes is also key to a number of diseases such as autoimmune syndromes and cancer.

Cell division

Time-based expression of proteins is important for driving pathways in which their functions are serially required, and a good example of this is the control of cell division. Every time a cell divides by mitosis the genomic DNA must be faithfully and completely duplicated and then equally distributed between the two daughter cells. Failure to accomplish this leads to mutations that, if retained, could carry the risk of cancer or, if lethal, to cell death. Cell death is an important means of tumour suppression but it can have ageing effects. The molecular interactions and regulation

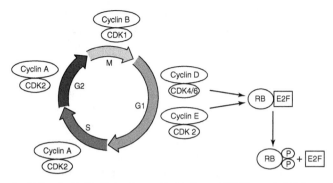

17. Cell cycle divided into phases, G1, S, G2, and M. The cycle is driven by the kinases (CDKs) that are activated by binding to sequentially expressed cyclins.

required for the cell cycle leading to cell division are therefore complex and tightly controlled. The cell cycle is a unidirectional process controlled by sequential and progressive expression of regulatory proteins called cyclins. For the sake of genomic fidelity there is a timely order for their expression and activity. Other key proteins interact, including those that provide strict checkpoints between stages of the cell cycle (see Figure 17).

The cell cycle is divided into phases; Gap1 or G1 is a growth phase in which the cell is prepared for DNA replication. The activity during G1 then initiates the transition from G1 to the synthetic or S phase. The cell is now committed to DNA duplication. After S phase the cell enters a second gap phase or G2 before finally dividing and the completion of mitosis (M). Cell cycle-regulated transcription therefore occurs in three main waves corresponding to the main transition points of the cell cycle, namely G1 to S, G2 to M, and M back to G1.

The temporally expressed regulatory proteins, called the cyclins, control the progress through cell division by activating enzymes called cyclin-dependent kinases (CDKs). Discovery of these key molecules led to the Nobel Prize in Physiology or Medicine being awarded to Paul Nurse and Timothy Hunt in 2001. The cyclin-activated CDKs act on target proteins to orchestrate transition through the phases of the cycle.

Cyclin D is the first to be produced, usually in response to growth stimuli. Cyclin D then binds to the kinases CDK4 and 6, activating them. The activated complex phosphorylates the retinoblastoma protein pRb, so named because lack of pRb underpins the formation of retinoblastoma tumours in the eye. The phosphorylated pRb releases the transcription factor E2F to which it was previously bound. E2F now travels into the nucleus and activates transcription of genes coding for the proteins that are necessary for cell division, including further cyclins, E, A, and B and the enzyme DNA polymerase.

Transition from G1 into S phase commits the cell to division and is therefore a very tightly controlled restriction point. Withdrawal of growth factors, insufficient nucleotides, or energy to complete DNA replication, or even a damaged template DNA, would compromise the process. Problems are therefore detected and the cell cycle halted by cell cycle inhibitors before the cell has become committed to DNA duplication. There are two families of cell cycle inhibitors, the CDK interacting protein kinase inhibitory protein (cip/kip) family and the Inhibitor of kinase 4/alternative reading frame (INK4/ARF) family. The cell cycle inhibitors inactivate the kinases that promote transition through the phases, thus halting the cell cycle. The INK4a protein or p16 binds to CDK4 preventing phosphorylation of pRb, and inhibiting the start of cell proliferation. The regulation of cell division provides an example of exquisite molecular control. The cell cycle can also be paused in S phase to allow time for DNA repairs to be carried out before cell division. The consequences of uncontrolled cell division are so catastrophic that evolution has provided complex checks and balances to maintain fidelity. The price of failure is apoptosis or programmed cell death.

Apoptosis and cell death

Apoptosis, named after the Greek for falling of autumnal leaves, is the name given to cell death that is programmed, as opposed to traumatic cell death or necrosis. Hitting your thumb with a hammer causes necrosis in which the cellular contents are spilled into the surrounding tissue fluid triggering a damaging and throbbing inflammatory reaction. In contrast, 50 to 70 billion cells die every day in a human adult by the controlled molecular process of apoptosis. This programmed method of cell death produces membrane-bound cellular fragments that are rapidly engulfed by surrounding cells. Excessive apoptosis can lead to premature ageing or to atrophy, whereas uncoupling of the apoptotic pathway from cellular damage signals underpins cancer formation.

Andrew Wyllie working with Alastair Currie and John Kerr in Aberdeen published the seminal paper describing the importance of apoptosis in 1972 (2). Sydney Brenner and John Sulston then went on to identify the genes involved in the process. Apoptosis can be triggered by activated external receptors on the plasma membrane or by internal sensors of stress, principally involving the p53 gene. Regardless of the initial stimulus, apoptotic pathways converge on the mitochondria causing loss of its membrane integrity and release of cytochrome C, a potent mediator of apoptosis. This activates a family of intracellular proteases called caspases that rapidly degrade cellular organelles and chromatin in a way that avoids inflammatory responses that are typically seen after cell necrosis.

Caspases are activated from their inactive pro-enzyme in a cascade that rapidly degrades the proteins in cellular organelles and chromatin, while endonucleases fragment the DNA, culminating in cell destruction. Members of the B cell lymphoma 2 or Bcl2

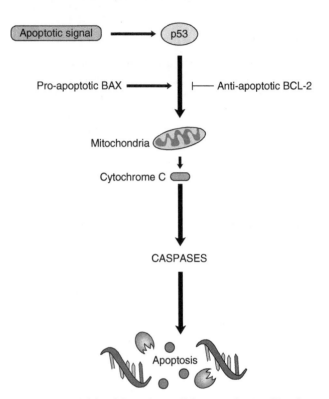

18. Apoptosis is initiated due to intracellular stress by signalling that releases cytochrome C from the mitochondria and activates a caspase cascade resulting in cell destruction.

family of proteins regulate the apoptotic pathway, some by inhibition others by promotion. Bcl2 is the original member of the family to be described and gets its name from the B cell lymphoma in which it was found aberrantly expressed. In normal tissues however Bcl2 promotes survival of healthy cells by inhibiting apoptosis. On the other hand Bcl2-associated X protein, or Bax, promotes cell death by binding to Bcl2 and inhibiting it. Activated Bax binds to the mitochondrial membrane making it leaky, resulting in the release of cytochrome C.

This initiates the suicidal cascade that results in apoptosis (see Figure 18). Bax is activated by p53 and executes p53's apoptotic function. Deregulated expression of Bcl2 or other members of the death inhibitory family cause unwarranted cell survival, a process common in certain cancers.

Regulation of gene expression is vital during life for determining responses both to internal stresses and to the environment. The processes involve complex layers of control from DNA and chromatin modification, transcriptional control, RNA processing, to protein production and activation. Failure to regulate gene expression has a high price resulting in disease and death.

Chapter 6
Genetic engineering

As we have seen, gene-cloning processes enable us to produce large amounts of a DNA sequence so that its function can be studied. These technologies can also be applied in medicine and agriculture to genetically engineer production of biological proteins or whole organisms with new or modified traits. At the heart of these applications is the capability to produce proteins from cloned genes in host cells. These proteins are called recombinant proteins as they are produced from recombinant DNA.

An advantage of recombinant proteins is that they can be produced relatively inexpensively in very large quantities. They can also be purified from the host in which they are cloned more easily than from original sources such as animal or human fluids, tissues, or plants. For example, the blood-clotting protein Factor VIII used to treat patients with haemophilia was previously derived from animal or human sources. However, both sources provide only low amounts and patients have been known to contract HIV and hepatitis C as a consequence of contaminating viral particles being purified at the same time. Thus recombinant proteins can also eliminate the risk of infectious diseases. Another benefit of recombinant proteins is that the gene can be modified, often through laboratory-based mutagenesis methods, to produce proteins with improved or novel functions.

Recombinant protein production is achieved by inserting the DNA coding for a protein of interest into a specially designed vector. These expression vectors carry features that enable gene cloning and additionally carry promoter and terminator sequences for transcription and translation in the foreign host. Ease of purification is made possible by tagging the protein with a marker for separating it from all of the other proteins produced in the host. For a protein to be useful it needs to be fully functional and this often requires newly synthesized proteins to be correctly folded and post-translational modifications to be made, such as the addition of sugar groups. This can be technically challenging, but despite the difficulties, a huge number of recombinant proteins have been made.

Recombinant pharmaceuticals

One key type of recombinant proteins is the therapeutic protein, also called biopharmaceuticals or biologics. There are many diseases that arise because a particular protein is either absent or a faulty protein is produced. Administering a correct version of that protein can treat these patients.

The first commercially available recombinant protein to be produced for medical use was human insulin to treat diabetes mellitus. This disease is caused by a lack of the protein hormone insulin that is required for the utilization of glucose in the body. A deficiency results in an accumulation of glucose in blood and urine, disturbing numerous cell functions, and can ultimately cause death. Insulin was previously purified from bovine and porcine pancreas. However, it was available only in low amounts, protein purification was difficult, and some patients' immune systems responded adversely. To overcome these problems the company Eli Lilly engineered recombinant human insulin protein. The human gene was isolated, cloned into a vector, and expressed in *E. coli* bacterial cells. The US Food and Drug Administration (FDA) approved the recombinant insulin for clinical use in 1982.

Since then over 300 protein-based recombinant pharmaceuticals have been licensed by the FDA and the European Medicines Agency (EMA) to treat a variety of disorders, including haemophilia, arthritis, heart attacks, and cancer, and many more are undergoing clinical trials.

Therapeutic proteins can be produced in bacterial cells but more often mammalian cells such as the Chinese hamster ovary cell line and human fibroblasts are used as these hosts are better able to produce fully functional human protein. However, using mammalian cells is extremely expensive and an alternative is to use live animals or plants. This is called molecular pharming and is an innovative way of producing large amounts of protein relatively cheaply. In animal pharming, sheep, goats, cows, or rabbits are often used. The first therapeutic protein produced in live animals was ATryn in goat's milk, a drug which prevents blood clotting. To achieve this, a DNA clone was constructed which carries the gene coding for the blood-clotting factor, attached to a promoter sequence. The promoter sequence is specifically selected to direct protein production where required; in the case of ATryn in the goat's milk. The DNA clone is injected into a fertilized egg removed from a recently mated female, where it integrates into the host genome. The fertilized egg is then implanted into the oviduct of a surrogate goat and the offspring that are born express the protein in their milk. The protein is extracted from the milk and formulated into a pharmaceutical product, for example, a tablet.

The production of cancer drugs in the whites of hens' eggs has also been recently reported. Here a gene coding for a particular protein derived from humans is combined with sequences from the hen genome so that protein production is directed to the egg whites.

In plant pharming, tobacco, rice, maize, potato, carrots, and tomatoes have all been used to produce therapeutic proteins. The process is similar to animal pharming in that the gene of interest combined with specific promoter sequences is transferred into plant cells using a vector. The promoter is selected to direct production to particular parts of the plant—leaves, seed, or fruit as desired. Molecular pharming is a controversial area of gene cloning as it relies on the production of transgenic organisms or plants, their genomes carrying DNA from an unrelated organism.

Monoclonal antibodies

Another class of proteins that can be engineered using gene-cloning technology is therapeutic antibodies. Antibodies are proteins that are produced by the body in response to viral, bacterial, or some other pathogenic infection. They defend the body by binding to specific proteins (antigens) found on the surface of pathogenic material and targeting it for destruction. Therapeutic antibodies are designed to be monoclonal, that is, they are engineered so that they are specific for a particular antigen to which they bind, to block the antigen's harmful effects. Previously, monoclonal antibodies were generated by fusing antibody-producing cells, the B-lymphocytes from an immunized mouse or rat with mouse tumour cells. This produced immortalized cells called hybridomas that secreted the desired antibody. However, rodent antibodies have limited use as they are not particularly stable in humans and can trigger allergic reactions. It is now possible to construct humanized antibodies using genetic engineering techniques. This involves replacing the rodent antibody coding DNA segments with human antibody coding regions and expressing the antibody in host cells. The first murine monoclonal antibody to be approved for use was Muromonab in 1986 to treat transplant rejection in patients. It works by binding to the CD3 receptor protein found on the surface of a type of immune cells called the T-cells. T-cells are involved in mounting an immune response to eliminate foreign material. When Muromonab binds to T-cells, the immune response is blocked and rejection of the transplant is prevented.

Bevacizumab, commonly known as Avastin™, is an example of an early humanized monoclonal antibody. It is used to treat colorectal, breast, and lung cancers. The antibody is designed to bind to the protein vascular endothelial growth factor (VEGF). This protein is expressed in tumour cells and causes new blood vessels to form so that the tumour can expand in size. By binding to VEGF, Avastin™ blocks blood vessel growth and limits tumour expansion. The experimental drug ZMapp developed to treat the Ebola virus disease is also a monoclonal antibody. It is composed of a combination of three monoclonal antibodies made from human and mouse genes, coding for proteins that target and inactivate the Ebola virus. The genes are inserted into vectors and manufactured in tobacco plants grown indoors in factory-like farms. Monoclonal antibodies are at the forefront of biological therapeutics as they are highly specific and tend not to induce major side effects.

Recombinant protein vaccines

Vaccines are biological preparations composed typically of a killed or inactivated form of a disease-causing organism. When injected, it stimulates the immune system to produce antibodies thus conferring immunity to that disease. Influenza, polio, and cholera vaccines use killed pathogens while in rubella, measles, and tuberculosis vaccines the pathogen is live but its infectious properties are disabled. An alternative to killed or disabled vaccines is to produce recombinant vaccines. This is possible if antibodies are produced in response to specific components (antigens) of the infectious agent and if the genes coding for those components are known. These coding sequences can be inserted into an expression vector to produce the recombinant protein. As the vaccine is composed of the antigen that triggers the immune response and not the whole pathogen it is less likely to trigger adverse reactions. A number of recombinant vaccines have now been approved for use. One is the vaccine against the Human Papilloma Virus generated by combining viral proteins from different strains to protect against genital warts and cervical cancer. Another is the hepatitis B vaccine, created using a protein found on the surface of the hepatitis B virus. The host cell for manufacture is typically a mammalian cell but expression in plants is also being trialled. Synthesis of recombinant protein conferring immunity against hepatitis B has been successful in banana, potato, tobacco, and carrots. When these 'edible vaccines' are consumed by an individual, the immune system builds up antibodies to fight the disease just like a traditional vaccine. Scientists are still working on how to make these vaccines more effective. They could represent a simple and cheap way of carrying out mass vaccination programmes, particularly in developing countries, overcoming the need for sterile needles, injections, and refrigeration.

Models of human disease

It is possible to generate an animal model for some human diseases to increase understanding of disease processes and test the efficacy of new treatments. Although a number of organisms can be used, the most common is mice since being mammals their genomes are similar to those of humans. Additional advantages are a short gestation period of about three weeks, a fairly large litter size, and a short lifespan of about two years.

A vast number of mouse models have been created for different diseases including cancers, cardiovascular disease, diabetes, neurodegenerative disorders, spinal cord injuries, AIDS, and obesity. These models are generated using genetic engineering techniques in which often an existing mouse gene is knocked out or knocked in. In a knock-out, the functional mouse gene is removed to mimic disease states that arise as a consequence of genes losing function. In a knock-in model, the mouse gene is replaced with a human gene carrying a specific nucleotide change and is used to mimic disease states in which mutated DNA produces a faulty protein.

Although mouse models are invaluable for biomedical research, there are limitations to their use. Some diseases such as Alzheimer's disease, Parkinson's disease, and psychiatric disorders like schizophrenia cannot be fully replicated in mice. In addition, some drugs show good results in mice but then fail to work or have adverse side effects in humans. An alternative to testing drugs in animals is currently being investigated at the Harvard Wyss Institute in the USA, using an organ-on-a-chip device. These chips, the size of a computer memory stick, contain human cells embedded on plastic and mimic the architecture and function of different organs such as lung, heart, skin, and intestines. Initial studies show that these can be used to study human disease processes and test drug responses. Work is ongoing to link the different organ chips together to develop a full human model system.

Gene therapy

In gene therapy the aim is to restore the function of a faulty gene by introducing a correct version of that gene. Gene therapy could potentially be used to treat inherited diseases arising through defects in a single gene such as cystic fibrosis but also more complex multi-gene disorders like cancer. The process sounds relatively simple in that a cloned gene is transferred into the cells of a patient. Once inside the cell, the protein encoded by the gene is produced and the defect is corrected. However, there are major hurdles that need to be overcome for gene therapy to be effective. One is the gene construct has to be delivered to the diseased cells or tissues. This can often be difficult with cells such as the brain which are hard to access compared to others like the eyes, skin, and muscles. Mammalian cells also have complex mechanisms that have evolved to prevent unwanted material such as foreign DNA getting in. Second, introduction of any genetic construct is likely to trigger the patient's immune response, which can be fatal in some instances. Another difficulty is that, once delivered, expression of the gene product has to be sustained to be effective.

One approach to delivering genes to the cells is to use genetically engineered viruses constructed so that most of the viral genome is deleted, removing the harmful capabilities of the virus. The therapeutic gene is inserted into these vectors, which then deliver the gene to the patient's cells. The viral vector can be modified in a way that makes it more effective at being taken up by the target cell, for example by the addition of surface proteins that recognize the desired cell.

Once inside the cell, some viral vectors such as the retroviruses integrate into the host genome (see Figure 19). This is an advantage as it provides long-lasting expression of the gene product. However, it also poses a safety risk, as there is little control over where the viral vector will insert into the patient's genome. If the insertion occurs within a coding gene, this may inactivate gene function. If it integrates close to transcriptional start sites, where promoters and enhancer sequences are located, inappropriate gene expression can occur. This was observed in early gene therapy trials during 2001–2 conducted in France, the USA, and the UK treating children with X-linked SCID (severe combined immunodeficiency) using a retroviral vector. Of the twenty patients treated, the immunodeficiency was corrected in seventeen patients, the first evidence that gene therapy can cure a life-threatening disease. However, five of these patients developed a T-cell leukemic disease between twenty-three and sixty-eight months later. This was due to the integration of the retrovirus at the transcriptional start site of a cancer-causing gene leading to enhanced gene expression. Since then, modified viral vectors have been investigated which direct integration at specific safe sites within the host genome.

Other viruses such as the adenoviruses do not integrate their DNA into the host genome but are maintained inside the cell as a separate entity. Similar to integrating viruses, the adenoviral vectors are engineered to remove portions of the viral genome so that they are not pathogenic to humans. However, adenoviral vectors do not

Gene transfer therapy

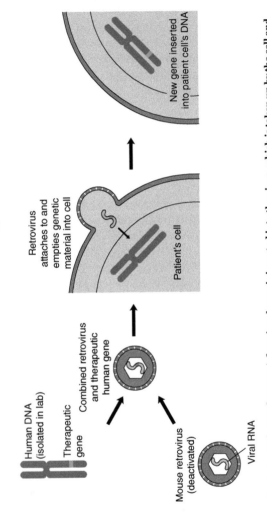

19. Retroviral gene therapy. A functional gene is inserted into the virus, which is taken up by the cell and integrates into the host genome.

provide long-term persistent expression of the transgene and are therefore not suitable for treating disorders where long-term expression is required. Adenoviruses can also provoke strong immune activation in the host, particularly in ill patients. This was highlighted by the Jesse Gelsinger case in 1999. In the trial in which Jesse took part, eighteen patients with Ornithine Transcarboxylase (OTC) deficiency, a genetic disorder which leads to neurological and gastrointestinal dysfunction, were administered gene therapy using adenoviral vectors. Jesse died of multiple organ failure just a few days after treatment as a consequence of the adenovirus producing a massive immune response.

An alternative to adenoviruses are the Adeno-associated viruses (AAVs). These viruses—unrelated to adenoviruses despite their name—are often used in gene therapy applications as they are non-infectious, induce only a minimal immune response, and can be engineered to integrate into the host genome, providing longer-term gene expression. However, AAVs can only carry a small gene insert and so are limited to use with genes that are of a small size.

Despite early disappointing results and tragic deaths, gene therapy has had some successes and the future looks positive.

Gendicine was the first gene therapy product approved for clinical use in humans in 2003 by China's State Food and Drug Administration (SFDA) for the treatment of head and neck squamous cell carcinoma (HNSCC). Gendicine is composed of a replication-deficient adenoviral vector and a functional human p53 tumour suppressor gene (*TP53*). *TP53* is mutated in 50 per cent of all tumours. Its role is to prevent abnormal cell proliferation. When its function is lost, damaged cells continue to proliferate inappropriately and tumours arise. By delivering a functional *TP53* to cells that lack it, diseased cells can be eliminated by the p53 protein triggering cell death. Clinical trials conducted by the company Shenzhen SiBiono GeneTech reported dramatic improvements in patients treated with Gendicine after just two months. These initial results have not been replicated in other clinical trials and China currently remains the only country to use Gendicine.

The first gene therapy to be approved for clinical use in Europe was Glybera, in 2012, to treat lipoprotein lipase enzyme deficiency. This is an inherited disorder that leads to high levels of blood fats causing abdominal pain and life-threatening inflammation of the pancreas. Glybera is composed of an AAV carrying a functional copy of the human lipoprotein lipase gene. It is injected into muscle cells for which the AAV vector has particular affinity. Once inside the cells, lipoprotein lipsase is produced, which helps break down fats and reduces pancreas inflammation.

An alternative delivery system to viruses is to package the DNA into liposomes that are then taken up by the cells. This is safer than using viruses as liposomes do not integrate into the host genome and are not very immunogenic. However, liposome uptake by the cells can be less efficient, resulting in lower expression of the gene. A recent success for non-viral gene therapy to treat cystic fibrosis has been reported. This is a genetic disease caused by mutations in the fibrosis transmembrane conductance regulator (*CFTR*) gene. The protein encoded by *CFTR* is responsible for controlling the movement of salts and water in and out of cells. When mutated, this control is lost and results in the lungs and digestive system in particular becoming blocked with mucus. The consequence is repeated infections and organ dysfunction. A phase 1 clinical trial has shown slightly improved lung function in patients receiving the *CFTR* gene complexed to a liposome, through a nebulizer. Earlier attempts to deliver functional *CFTR* gene to cells using viral delivery methods have largely been ineffective due to poor uptake by cells. Currently there are over 2,000 gene therapy products in clinical trials, mainly to treat cancer but also genetic disorders and cardiovascular diseases.

Genetically modified foods

Genetic engineering technology can be applied to plants not only to manufacture therapeutic drugs but also to engineer plants that have increased nutritional value, and that repel weed killers and are resistant to droughts and infections. Farmers have for centuries selected plants for desirable traits—the most resilient, the tallest, and the tastiest—to breed subsequent generations. In conventional breeding methods, the DNA from the parents, when cross-bred, combines randomly. The appearance of the selected trait is therefore correspondingly unpredictable. In contrast, genetic engineering techniques allow targeted genetic changes to be made to the genome, leading to more reliable production of plants with the desired trait.

The first genetically modified food to be granted a licence for human consumption was a tomato called Flavr Savr in 1994 produced by the company Calgene. The tomato was engineered to delay ripening and hence extend the time it takes for it to soften and spoil. This was achieved by knocking out the expression of the gene coding for the enzyme endo-polygalacturonase, which is responsible for fruit softening. This product was taken off the market in 1997 amidst public safety concerns about genetically modified foods. More recently, scientists in the UK have engineered 'purple' tomatoes. These tomatoes produce high levels of a chemical called anthacyanin, which gives the tomatoes their purple colour. Anthacyanin is an anti-oxidant naturally found in fruits such as blueberries and blackberries. Its production is triggered in tomatoes by transferring genes from a snapdragon plant into the tomato genome. Studies have shown that anthacyanin can reduce the incidence of cancer and improve cardiovascular function and health. The aim would be to apply the purple tomatoes to food that people eat more frequently and are cheaper to purchase such as ketchup. This product is currently being tested for positive health benefits in humans.

The company AquaBounty Technologies is involved in the production of salmon that have been engineered to grow faster and reach their full size quicker. These salmon are engineered by combining a growth hormone coding gene from Pacific salmon with promoter sequences from an eel-like fish. This modification allows the growth hormone to be produced within the fish all year round instead of only a few months. Consequently, the full size is reached in half the time—eighteen months instead of three years—and can be marketed earlier. The FDA has now approved the genetically modified salmon for human consumption.

Herbicide-tolerant and insect-resistant crops

Two major reasons for crop loss are damage by weeds and insects. Crops are regularly sprayed with herbicides and insecticides to reduce crop loss, but these are often non-specific, damaging the crops being cultivated and in some cases causing harm to humans and to other organisms in the local environment. To overcome these issues and improve yields, crops have been established that are herbicide-tolerant or insect-resistant or carry both traits within the same crop at the same time.

Herbicide-tolerant crops are engineered so that when particular herbicides are applied to the field, the weeds are killed but the crop remains intact. The most common herbicide-tolerant crop commercially available is 'Roundup Ready' produced by Monsanto. Roundup Ready crops carry a gene encoding an enzyme that makes it resistant to applications of the broad-spectrum weed-killer glyphosate.

To produce insect-resistant crops, the gene encoding a toxic protein from the bacteria *Bacillus thuringiensis*, known as Bt, is transferred into the crop plant. The crop produces Bt protein that is toxic to certain insects but harmless to other mammals including humans, birds, and fish.

A project has been launched recently in the UK to generate a genetically modified potato that carries a wider range of traits. It will be resistant to late potato blight and to infection by potato cyst nematodes, diseases that cause substantial potato loss to farmers. It will also contain much lower levels of the naturally occurring chemicals, asparagine and reducing sugars. When cooked at high temperatures, these chemicals form acrylamide. Acrylamide in food has been linked to cancer and hence reducing consumption could potentially decrease the risk of cancer formation. The potato will also be less prone to bruising as it will have the gene coding for the enzyme polyphenol oxidase switched off. To generate this 'mega-potato', three genes conferring resistance to late potato blight and two genes conferring resistance to nematodes will be added to the potato genome. In addition, three genes that produce asparagine, the reducing sugars, and polyphenol oxidase will be switched off. Although still some years away from the final crop, if successful, the potato could increase yields, prevent large quantities of pesticides and fungicides being sprayed on agricultural land, and lead to the production of a healthier food.

Crops with improved nutritional value

Nearly 200 million people across the globe are currently estimated to suffer from vitamin A deficiency. Vitamin A deficiency affects pregnant women and children in particular and can lead to blindness. Supplementation of Vitamin A through the diet can alleviate its effects. One way of supplementing would be through consumption of Golden Rice, so called because of its intense yellow colour. Golden Rice is produced through the insertion of two genes into the rice genome; one derived from the daffodil and a second from a soil bacterium. These genes code for enzymes involved in the production of beta-carotene, a precursor to Vitamin A production. When consumed by an individual, beta-carotene is converted to Vitamin A in the gut. Details of how the strain is engineered were first published in the journal *Science* in 2000 and were seen as a significant breakthrough at that time.

Since then a number of studies have shown that Golden Rice consumption is as good as vitamin A supplements in children and better than natural beta-carotene found in spinach. The product has yet to be approved commercially.

Addressing GM food concerns

The first GM food was approved over twenty years ago; however, a wide range of GM food is still not available on the market and they remain a source of public debate internationally. One concern is that GM food consumption may pose a risk to human health through transfer of antibiotic resistance genes. These genes are used as selectable markers during genetic engineering processes and could be passed to humans through consumption of these foods. Another concern is the impact on the environment. The increased use of herbicide-tolerant crops has led to the development of 'super-weeds' that are resistant to herbicide applications. Resistance by some insects to Bt transgenic crops is emerging and a decline or elimination of natural wildlife has also been observed. In the USA, milkweed found in fields is a key food source for butterflies. Use of GM crops has eradicated the milkweed and in doing so eliminated the butterfly's food source. There are also concerns about non-GM crops being contaminated by pollen drifting from GM crops and producing genetically modified offspring. To limit the risks, a number of technological improvements are now applied. One is replacing antibiotic resistance marker genes with safer plant-derived markers. Another is to use promoter sequences that switch on protein expression only in response to a particular stimulus such as a chemical or stress factor. These inducible promoters not only restrict protein production to specific parts of the plant but also limit protein production temporally. Genes can also be inserted more precisely at specific sites of the genome using sophisticated genome-editing technologies, overcoming the random integration seen in earlier experiments. Increased research and further improvements in recombinant DNA technology should continue to minimize the potential risks of GM foods.

Chapter 7
Molecular biology in the clinic

Environmental agents can cause genetic and epigenetic changes to DNA, the consequences of which lead to deregulation of cellular processes and pathways that cause disease. Genetic variation can either be inherited if acquired through the germline or non-heritable when the DNA changes occur in somatic (body) cells. Some mutations are retained in a population if they confer an advantage when present with a fully functional wild type gene (heterozygous). These mutations only cause disease when both copies of the gene are mutant (homozygous). Often the inherited mutations have been selected for their beneficial roles in our ancestors; for example, the cystic fibrosis gene may have reduced the risk of death due to cholera.

Two key contemporary areas of clinical research in particular have benefited from an improved knowledge of their molecular basis; they are ageing and cancer. We are now better able to predict disease risk and design drugs that have higher clinical efficacy by targeting specific molecular pathways.

The molecular biology of ageing

We live in an ageing society and if present trends continue a fifth of the global population will be over the age of 60 by 2050. This will lead to a sharp increase in the health burden from chronic diseases of ageing, dubbed the 'Silver Tsunami'. Normal ageing results in part from exhaustion of stem cells, the cells that reside in most organs to replenish damaged tissue. As we age DNA damage accumulates and this eventually causes the cells to enter a permanent non-dividing state called senescence. This protective ploy however has its downside as it limits our lifespan. When too many stem cells are senescent the body is compromised in its capacity to renew worn-out tissue, causing the effects of ageing. This has a knock-on effect of poor intercellular communication, mitochondrial dysfunction, and loss of protein balance (proteostasis). Low levels of chronic inflammation also increase with ageing and could be the trigger for changes associated with many age-related disorders.

Dementias

Ageing is probably most associated with debilitating loss of brain function and associated personality disorders, which were first recorded by ancient Greek and Roman philosophers including Pythagoras. Dementia is a term for a broad category of brain pathologies including Parkinson's and Alzheimer's diseases (commonly shortened to Alzheimer's). Alzheimer's was described by and named after the German psychiatrist and pathologist Alois Alzheimer in 1901. Alzheimer's is the most common form of dementia for which there is currently no cure. It comprises up to 70 per cent of dementia cases and probably affects at least 7 per cent of the population. Prevention or delay may be possible by lifestyle choices but until we understand the molecular nature of the disorder, advice can only be of limited value. Alzheimer's is characterized by the presence in the brain of misfolded proteins that are resistant to normal enzymatic recycling. This class of disorder comes under the name proteopathy. The characteristic protease-resistant extracellular plaques that disrupt neuronal functions are composed of the protein amyloid. This finding led to the amyloid hypothesis of dementia. A gene on chromosome 21 codes for amyloid precursor protein (APP) and individuals with an extra copy of chromosome 21 (associated with Down's syndrome) suffer from early onset Alzheimer's. Another protein called tau also identifies Alzheimer's. Abnormally modified tau forms neurofibrillary tangles of twisted protein fibres within a nerve cell body, causing it to malfunction and die.

Recent genome-wide association studies (GWAS) have found genes that confer risk of dementia and these suggest mechanisms to explain the molecular pathology of the disease. Parkinson's disease affects up to 3 in every 1,000 people in the western world, including the celebrities Michael J. Fox and Muhammad Ali. The key to understanding Parkinson's disease may lie with the discovery of its association with a new protein called Tigar that, when overactive, kills nerve cells. Seven to ten million people worldwide are living with Parkinson's disease and the causes remain unknown. Therapeutic targeting of Tigar may halt the progress of this devastating disease.

Diseases of premature ageing

Much has been learnt about ageing from the molecular biology of inherited disorders of premature ageing, the progerias. The word progeria comes from the Greek meaning premature old age. The condition is extremely rare since sufferers do not live long enough to reproduce, so each case of progeria must be due to a new mutation. Symptoms arise in early infancy with a failure to thrive and noticeable ageing of the skin. Atherosclerosis and cardiovascular disease occur in childhood. Facial wrinkles and hair loss follow, along with loss of body fat and muscle, and stiff joints; all symptoms usually only seen in old age. The cause is a mutant form of the lamin A protein called progerin. The mutant lamin A impairs the structure of the nuclear envelope or lamina and therefore its functions, such as DNA damage repair and chromatin function. Progerin may play a role in normal human ageing since its production is activated in senescent cells. Other accelerated ageing disorders such as Werner's syndrome,

Xeroderma pigmentosa, or Cockayne's syndrome are also due to diminished DNA repair.

Pathways leading to old age

There has been dramatic progress in ageing research using yeast and invertebrates, leading to the discovery of more 'ageing genes' and their pathways. These findings can be extrapolated to humans since longevity pathways are evolutionarily conserved between species. The major pathways known to influence ageing have a common theme, that of sensing and metabolizing nutrients.

The story of rapamycin starts in the 1960s with a soil sample from Rapa Nui (Easter Island) that proved to have an activity capable of killing cells. This activity was later attributed to a small molecule, rapamycin, which was produced by soil bacteria. The field was advanced by identification of the *m*ammalian *T*arget *O*f *R*apamycin, aptly named mTOR. mTOR acts as a molecular sensor that integrates growth stimuli with nutrient and oxygen availability. Small molecules such as rapamycin that reduce mTOR signalling act in a similar way to severe dietary restriction in slowing the ageing process in organisms such as yeast and worms. Aberrant functions usually associated with ageing also occur in a range of diseases and rapamycin has been the focus of intense research both from academics and pharmaceutical companies. Rapamycin and its derivatives (rapalogs) have been involved in clinical trials on reducing age-related pathologies such as Alzheimer's, type II diabetes, obesity, and certain cancers. It is unlikely however that these or any other drug will be clinically approved to increase normal longevity since ageing is a normal process, not a disease. However a deregulated mTOR pathway is implicated in diseases associated with ageing and current research is being translated into health improvements and novel therapies that alleviate or prevent chronic disease rather than being targeted directly at extending lifespan. Most people would not wish for a prolonged lifespan without concomitant improvements in health.

Telomeres and diseases of ageing

Another major ageing pathway is telomere maintenance. The ends of linear chromosomes have specialized caps, analogous to the protective tips on shoelaces (aglets) that stabilize them to prevent damage. These chromosome caps are composed of a protein complex called shelterin bound to the repetitive DNA sequences (TTAGGG in humans). Thousands of copies of this sequence end in a single strand overhang that is tucked back into the chromosome where it is held in place by proteins. This is rather like anchoring the end of a thread when darning or weaving. The whole complex is known as a telomere from the Greek for 'end part'. Telomeres prevent chromosome ends from being recognized as damaged DNA and eliciting the double-strand break repair mechanism that would lead to chromosomes 'sticking' together or rearranging in a way that causes loss of genomic integrity.

Telomeres shorten with each cell division, a necessary evil caused by the failure of the DNA polymerase to complete a lagging strand replication. This is known as the 'end-replication problem'; and has been proposed as a biological clock that counts down lifespan from birth. Every time a cell divides it loses a small amount of DNA from its telomeres. Eventually the telomeres reach a critical length, triggering senescence and putting the cell into a non-replicative state. Thus telomeres confer a finite lifespan on a cell unless the telomeric DNA can be topped up in some way. Interestingly telomeric extension to levels seen in a neonate only seems fully possible in germ cells but it does occur to a limited extent in stem cells. The most common means of telomere repair is the addition of telomere repeats to the ends of chromosomes by the enzyme telomerase. An alternative mechanism called Alternative Lengthening of Telomeres or ALT has been described for some cell types. Telomere attrition is a hallmark of ageing and studies have established an association between shorter telomere length (TL) and the risk of various common age-related ailments such as cardiovascular disease, type II diabetes, and cancer.

Telomere loss is accelerated by known determinants of ill health, including chronic stress, smoking, excessive alcohol consumption, and obesity. Inherited genetic variations act in addition to these environmental factors to determine telomere length. This has been highlighted by recent genome-wide association studies that have revealed associations between telomere length and inherited variations in genes directly involved in telomere maintenance. These include genes that code for the protein and RNA components of the enzyme telomerase, TERT and TERC respectively, and other genes involved in telomere maintenance such as *CTC1*. Patients with mutant *CTC1* suffer from Coats plus disorder, a rare syndrome associated with low life expectancy and short TL. This supports a causal role for telomere biology in human ageing.

In population studies TL determination is carried out on white blood cells (leucocytes) since these are easily obtained. Lifestyle and genetic variations associated with short blood leucocyte TL (LTL) predict increased risks for atherosclerosis and diminished longevity. The relationship between TL and cancer appears complex. Short TL may be a risk factor for some cancers but for melanoma the opposite seems to be true, with long TL being the danger. The reason is uncertain but long TL may signify upregulation of telomere maintenance, a factor preventing normal senescence in melanocytes, thus promoting survival of pre-cancerous cells.

Cancer

Understanding the causes of cancer is paramount to its prevention and treatment. As long ago as the early 18th century lumps and polyps were surgically removed to reduce deaths from cancer.

Later that century Percival Potts, the first scientist to describe an environmental carcinogen, noticed an association between scrotal cancer and exposure to soot. He recommended that chimney sweep boys wear protective clothing rather than work naked. The latter was the normal practice to prevent soiling their only set of clothes. For the most part though, cancer and its origins were not comprehended until the advent of molecular biology.

Cancer is not a single disease but a range of diseases caused by abnormal growth and survival of cells that have the capacity to spread. An adult human is composed of 10^{14} cells, which remains relatively constant. This is achieved by a tight balance between cell proliferation (replication) and cell death. The body can increase in size due to cells getting bigger, for example fat cells in obesity or muscle cells due to exercise (hypertrophy), but rarely do increases in cell number (hyperplasia) persist. If cell birth exceeds cell death, new growth or, to give its Greek name, neoplasia would result. Tumours, from the Latin this time, are swellings composed of neoplastic cells. 'New growths' or neoplasms can be 'benign' if excessive cell proliferation is localized, or malignant if they invade into surrounding structures. Malignant tumours were given the name 'cancer' due to the crab-like appearance of their invasive projections. The most aggressive cancers spread to adjacent and distant organs by a process known as metastasis.

Cancer can affect any organ or tissue that contains dividing cells and develops due to uncontrolled cell proliferation. A normal cell is subject to a set of complicated molecular controls that limits inappropriate cell proliferation either through the cell undergoing cell cycle arrest or by inducing apoptotic cell death. In cancer cells, these controls break down leading to unrestrained proliferation. To support this unbridled growth, additional energy is required and thus cancer cells reprogramme their metabolic pathways to acquire this. As cancer cells are not normal, they should be recognized as such by the body's immune system and destroyed. However, by adopting various strategies they avoid destruction and may even hijack the immune system for their own advantage. Another important feature of cancer cells is their capability to grow new blood vessels, a process called neo-angiogenesis. The stimulus for angiogenesis is oxygen starvation or hypoxia, which arises in regions of the tumour displaced a short distance away from a blood vessel. Hypoxia would normally trigger cell death but recruitment of new blood vessels enables the tumour to continue to expand in size.

Eventually some cells migrate from the primary tumour, establishing secondary tumours at new sites. For example, cancerous cells in an epithelium lose contact with each other and take on migratory and invasive properties. One of the early stages in the acquisition of an invasive phenotype is epithelial-mesenchymal transition (EMT). Epithelial cells form skin and membranes and for this they have a strict polarity (a top and a bottom) and are bound in position by close connections with adjacent cells. Mesenchymal cells on the other hand are loosely associated, have motility, and lack polarization. The transition between epithelial and mesenchymal cells is a normal process during embryogenesis and wound healing but is deregulated in cancer cells. EMT involves transcriptional reprogramming in which epithelial structural proteins are lost and mesenchymal ones acquired. This facilitates invasion of a tumour into surrounding tissues.

Cancer is a genetic disease

Cancer is a genetic disease but mostly not inherited from the parents. Normal cells evolve to become cancer cells by acquiring successive mutations in cancer-related genes. There are two main classes of cancer genes, the proto-oncogenes and the tumour suppressor genes. The proto-oncogenes code for protein products that promote cell proliferation. These products are often growth factors and their receptors that act within signalling pathways to promote cell proliferation. A mutation in a proto-oncogene changes it to an 'oncogene', a term first coined in 1969 by George Todaro and Robert Heubner. Deregulated cell proliferation per se is not oncogenic as the tumour suppressor genes counteract it by inducing cell senescence or cell death. This makes cancer a relatively rare possibility unless mutations arise in the very genes that form the tumour suppression system. Therefore cancer arises when a mutation occurs that evades the normal controls and allows survival and proliferation of deregulated cells.

For over a century it has been recognized that viral and bacterial infections are risk factors for cancer. At least 15 per cent of cancers are attributable to infectious agents, examples being HPV and cervical cancer, *H. pylori* and gastric cancer, and also hepatitis B or C and liver cancer. Sometimes the viral genes become inserted into the human genome where they are continually expressed. Many of these viral gene products subvert the main tumour suppression pathways, in particular p53 and pRb—the belt and braces of proliferation control. It was the study of an oncogenic protein from a monkey virus (SV40) that led to the discovery of the master tumour suppressor gene *TP53* that codes for the protein p53.

TP53—Guardian of the Genome

One gene above all others is associated with cancer suppression and that is *TP53*. David Lane in 1992 dubbed it 'The Guardian of the Genome', Karen Vousden 'The Death Star' in 2000, and DePinho 'Good cop Bad cop' in 2002. It was Molecule of the Year in 1993 as it became recognized that approximately half of all human cancers carry a mutated *TP53* and in many more, p53 is deregulated.

p53 has many functions ascribed to it including apoptosis, genomic stability, and inhibition of the formation of new blood vessels or neo-angiogenesis. It is activated in response to various stress signals such as cells exposed to low nutrients, low oxygen, and to oncogene activation. It acts mainly in the nucleus as a transcription factor switching on genes required for homeostatic functions. In the cytoplasm, p53 has also been described as having a more direct role in activating apoptosis at the mitochondria. As a first line transducer of the stress response, p53 must always be present, even in an unstressed cell. Transcription and translation of p53 would be far too lengthy a process for a vanguard defender of homeostasis, so p53 is held at low levels through continuous synthesis and degradation, ready to be activated. Activation is caused by loss of the inhibitor protein MDM2 that normally promotes the destruction of p53. p53 plays a key role in eliminating cells that have either acquired activating oncogenes or excessive genomic damage. Thus mutations in the *TP53* gene allow cancer cells to survive and divide further by escaping cell death (see Figure 20).

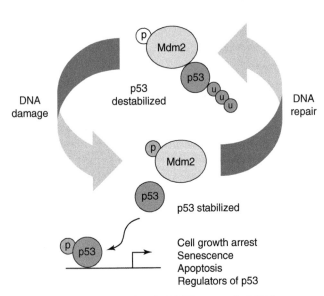

20. p53 levels are controlled by the inhibitor protein, MDM2.

A mutant p53 not only lacks the tumour suppressor functions of the normal or wild type protein but in many cases it also takes on the role of an oncogene. In fact p53 was first classed as an oncogene until it was discovered that the tumour-promoting properties found were due to a mutated protein. Mutant p53 can gain functions that increase its stability, abrogate the function of a wild type protein, and actively promote tumour growth.

Naturally occurring polymorphisms in p53 can affect its function. Codon 72 in *TP53* either codes for the amino acid proline or arginine. This difference affects the structure of p53 and its function, with arginine at 72 reported to be more efficient at inducing apoptosis than the ancestral proline variant. In the northern hemisphere distinct geographic differences in the frequency of proline and arginine have been observed. The proline variant shows a north–south gradient, with a frequency of only 17 per cent in Scandinavian Laplanders but up to 63 per cent in Nigerians (see Figure 21). In western Europe the arginine allele is the most common with frequencies up to 83 per cent reported. The proline on the other hand is frequent in African Americans. These latitude-dependent variations have been suggested to be due to selection related to winter temperature or to UV radiation. Pale skin is necessary where annual UV is low in order to make vitamin D and the arginine variant being more apoptotic may be required for removing UV damaged cells in pale-skinned individuals. Alternative theories include the arginine variant being more efficient at inducing tanning; aiding fertility in Caucasians or promoting tolerance to a high fat diet.

Germ line mutations in *TP53* can also occur, giving rise to the Li–Fraumeni syndrome named after the two doctors who first identified the heritable link between a range of tumours arising in breast, brain, and white blood cells (leukaemia). Overall 5–10 per cent of cancers occur due to inherited or germ line mutations that are passed from parent to offspring. Many of these genes code for DNA repair enzymes such as the susceptibility

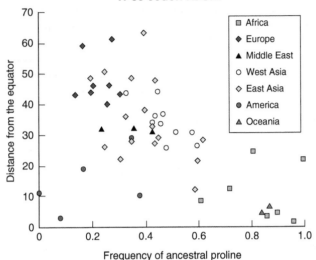

TP53 codon 72 SNP

Legend:
- □ Africa
- ◆ Europe
- ▲ Middle East
- ○ West Asia
- ◇ East Asia
- ● America
- ▲ Oceania

x-axis: Frequency of ancestral proline

y-axis: Distance from the equator

21. **Geographical distribution of the *TP53* codon 72 variants or SNPs. The ancestral variant with proline is found nearer the equator with the arginine version more common nearer the poles.**

gene *breast cancer 1 early onset* or *BRCA1*. Germ line mutations in enzymes involved in the repair of mismatched bases in the DNA underlie some forms of inherited colon cancer.

The vast majority of cancer mutations are not inherited; instead they are sporadic with mutations arising in somatic cells. Mutations arise due to uncorrected errors in replication but also due to exposure to environmental agents such as the chemical carcinogens in tobacco smoke or by UV radiation. If mutations remain uncorrected and if they occur in dividing cells the damage is transmitted to the daughter cells. Failure to correct errors in DNA creates instability within the genome, and additional mutations accumulate over time. This increases the chances of the mutant cell progressing to cancer. Thus cancer development is complex, involving interactions between genes and the environment. As multiple mutations are required in different biological pathways, the disease usually takes time to progress and is therefore commonly associated with older individuals.

Telomeres and cancer cells

Normal human cells have a finite lifespan, at the end of which they cease to divide and enter a state or replicative senescence; a process in itself that is tumour suppressive. The reason for this is telomere erosion. p53 detects the short eroded telomeres and prevents further cell division. In contrast tumour cells can have the capacity to divide forever as seen in the immortal cell lines such as HeLa that have been derived from human tumours. Tumour cells must therefore acquire a persistent telomere maintenance mechanism and for most cancers this comes in the form of telomerase, which adds telomere repeats to the end of chromosomes. Some tumours become immortal in the absence of telomerase and the alternative process, ALT, repairs their telomeres.

Non-coding RNAs and cancer

Non-coding RNAs are now recognized as important players in cancer formation. Prostate cancer is one of the leading causes of cancer in men and at least six ncRNAs are deregulated in prostate cancer, some of which are driven by male hormones or androgens. Cancer-specific chromosomal translocations put male hormone response elements next to ncRNA genes causing their deregulation. This results in a cascade of cellular events leading to carcinogenesis and tumour progression. ncRNAs offer an attractive possibility as markers for detection and risk stratification of prostate cancer and may also make excellent therapeutic targets. In the future RNA-based therapies may become a widespread option for cancer treatment.

Cancer stem cells

Not all the cells comprising the mass of a tumour possess the capacity to regenerate it. There is evidence to suggest that cancers are maintained by a minority population of cells within the tumour bulk called cancer stem cells (CSCs). CSCs, like their normal counterparts, are somewhat resistant to antiproliferative therapies. The reason why current cancer therapies often fail to eradicate the disease is that the CSCs survive current DNA damaging treatments and repopulate the tumour. Even modern targeted treatments may ultimately fail due to the presence of CSCs, which must therefore become the targets for future therapies.

New-targeted drugs

Recent large-scale cancer genome sequencing projects have provided a vast amount of information on the genetic changes associated with cancer. The significance of the minority cancer stem cell population must however be taken into account when analysing this data. Microarray analysis comparing expression profiles of cancer cells and non-cancer cells has provided some useful insights. As we now have an improved understanding of the molecular pathways of cancer, we are in a better position to devise treatments. Until recently, treatment of cancer was limited to surgical removal of tumours, chemotherapy, and radiotherapy. While these are still in use, the newer approach is to target specific molecules unique to cancer. This is a challenge given the number and complexity of the biological pathways involved. However, the aim is to disable the function of growth-promoting oncogenic proteins or reactivate the function of tumour suppressors.

One of the first targeted drugs to be developed was Imatinib or Gleevec^{TM}, a designer drug marketed by Novartis for the treatment of a type of leukaemia called chronic myelogenous leukaemia (CML). CML starts as a prolonged chronic phase, with high white blood cell counts driven by a characteristic chromosomal translocation known as Philadelphia chromosome (Ph). The Philadelphia chromosome generates the oncogenic enzyme BCR-Abl. The BCR or breakpoint cluster region on chromosome 22 becomes joined to the Abl kinase from chromosome 9q. This removes the normal control of Abl expression that now becomes stuck in the 'on' position. The continually expressed Abl enzyme activates a proliferation pathway in the Ph+ leukaemia CML cells. Imatinib is the paradigm for targeted cancer therapeutics designed with the aid of crystallographic models to fit into the active site of this fusion protein thus blocking the activity of the mutant enzyme (see Figure 22). The treatment has very low side effects since it specifically hits the cancer cells harbouring the mutant enzyme and spares normal cells. This gives patients relief from the morbidity of the disease and an extended survival time but unfortunately it is not curative, as it doesn't kill the cancer stem cells. Ph+ leukaemia cells persist as detected by

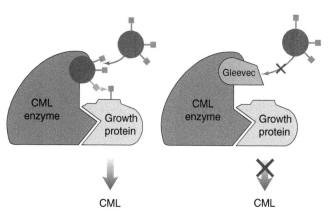

22. Gleevec™ is a drug specifically designed to inhibit the mutant growth protein or kinase formed by fusion of BCR and ABL in chronic myeloid leukaemia cells.

RT-PCR and the patient requires continuous treatment. Eventually additional mutations arise in CSCs that confer drug resistance and CML transforms into a fatal phase, indistinguishable from acute leukaemia. Some of these new mutations have been studied and new designer drugs trialled.

Chapter 8
Molecular forensics

A key challenge for molecular biology is to use research outcomes to meet the demands of modern society. Molecular markers are having a substantial impact; from bench to boardroom or even courtroom. DNA profiling plays a role in helping to solve crimes and miscarriages of justice. Although 99.5 per cent of the human DNA sequence is the same for everyone, there are small regions of variability that are specific to each individual giving each person a unique DNA profile, or fingerprint. DNA profile applications are increasing and are now used to identify food counterfeits and contamination of food.

DNA profiling—using the unique sequence in our genomes

In 1977 Alec Jeffreys, now Sir Alec Jeffreys FRS, applied molecular biology techniques to analyse inherited variation in human genes. The method of choice at that time was Restriction Fragment Length Polymorphism (RFLP). This involved using restriction enzymes that are sequence specific to cut DNA into shorter pieces. The DNA fragments are then separated by size on a gel before transferring them to a membrane. Once attached to the membrane or Southern blot, the DNA is probed with a labelled sequence from the gene of interest. A single mutation or base difference in a gene would either create or destroy a restriction site on the DNA, producing a fragment of a shorter or longer length respectively. On a gel the fragments migrate at different speeds and therefore appear as bands of different size on the Southern blot. However Jeffreys found that human DNA within genes did not show much variation, so he switched to using DNA from in-between genes. This work led to the development of DNA fingerprinting. One morning in 1984 Jeffreys developed a blot in his Leicester University darkroom. The blot was of DNA from his research assistant and family members and what he saw for each DNA sample was an array of bands rather like a barcode. Jeffreys had discovered the first example of a human 'minisatellite' and he immediately recognized the significance of his finding. Minisatellites are sequences of DNA consisting of around thirty base pairs that are repeated tens to hundreds of times. The number of these variable tandem repeats is heritable and extraordinarily variable between individuals. What Jeffreys and his team now had was a mechanism for displaying large numbers of minisatellites that would provide a highly individual pattern or fingerprint of human DNA. For example, DNA fingerprinting allows easy comparisons to be made between DNA left at a crime scene and that of several potential suspects (see Figure 23). These fingerprints can be used not only for forensics, but also paternity and immigration disputes, to name but a few applications.

The potential of this technology came to the attention of a lawyer who in 1985 was defending a young boy in danger of being deported and therefore separated from his alleged family in Britain. It was an opportunity to put the molecular fingerprinting technology to practical use. DNA was taken from the boy, his parents, and other family members and analysed. The pattern of bands on the gel showed that the boy was the genuine son of the British parents and therefore he was not deported to Africa. After that, DNA fingerprinting led to a change in the immigration laws.

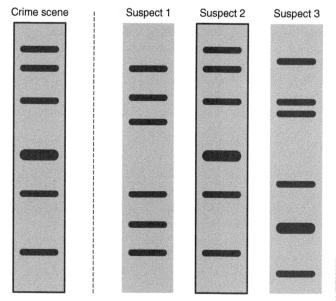

| Crime scene | Suspect 1 | Suspect 2 | Suspect 3 |

23. DNA fingerprinting is used in forensic science to identify matches between DNA from a crime scene with that of a suspect.

There are about 10 million different sites at which people can vary in their DNA sequence within the 3 billion bases in our DNA. Professor Jeffreys and his team worked on refining their technique to make use of a larger number of these variations, thus making it suitable for databasing. This was now called 'DNA profiling' and it was about to change the face of criminology forever. A few, but highly variable sequences or minisatellites are chosen for DNA profiling. These give a highly sensitive procedure suitable for use with small amounts of body fluids and Jeffreys tested this out in the laboratory by 'spotting' blood on several surfaces, and then retrieving and analysing it. DNA profiling was soon to be tested in court. Two young girls had been raped and murdered, one in 1983 and the other in 1986. Was this the work of a serial killer, or two unrelated cases? The police arrested a man and obtained a confession to the second murder. The two cases seemed so similar that the police thought their suspect must have committed both murders. Jeffreys analysed the forensic samples for DNA profiles and proved that the same man had raped both girls. However, it was not the suspect, who had obviously made a false confession. The police now had to find their man and, armed with the DNA profile of the criminal, set out to test blood samples from 5,000 men in the area. The match was found, and the suspect sentenced to life in prison.

From minisatellites to microsatellite databases

Forensic scientists the world over wanted to set up databases of DNA profiles to aid investigations. There was a demand for a simplified and more sensitive technique. The answer was to amplify the variable regions by PCR. However, the minisatellite probes used up to now were too long for PCR technology so even shorter sequences called microsatellite repeats were used. Each marker or microsatellite is a short tandem repeat (STR) of two to five base pairs of DNA sequence. A single STR will be shared by up to 20 per cent of the population but by using a dozen or so identification markers in a profile, the error is minuscule. A DNA profile would be recorded as one or two alleles or differences at each of at least fifteen STRs. The US DNA database is called CODIS (Combined DNA Index System) and it contains over ten and half million samples of genomic DNA profiles from known convicted offenders and unidentified DNA left at crime scenes. The database is searched for an exact match with a DNA profile left at a crime scene. The US requires a minimum of thirteen STRs that match to make an identification, so a minimum of fifteen STRs are analysed to ensure at least thirteen are useful. If the crime-scene material differs in any of those thirteen places it is not counted as a complete match and the police do not have their suspect. Very close matches can however point to close relatives.

In the mid-1980s, a serial killer nicknamed 'Grim Sleeper' murdered at least ten women in the Los Angeles area. He eluded capture for nearly twenty-five years but was finally apprehended as a result of DNA databasing. At first no match was found in the Californian database but a couple of years later a DNA sample came in from a gun felon that had a remarkable similarity to the Grim Sleeper's DNA. The most likely explanation was that it came from a close member of Grim Sleeper's family. The gun felon was too young to have carried out murder in the 1980s so suspicion focused on an older relative, possibly his father. A DNA sample was obtained from the father and a match obtained with the DNA from crime-scene evidence collected long ago. The Grim Sleeper was finally arrested.

DNA profiles are being used for countless legal and medical benefits. They are being used to counteract the enormous international industry of human trafficking. Children who are suspected of being trafficked can be tested and reunited with their parents. For example, an organization based in Spain called DNA-PROKIDS has used DNA profiling to identify more than 700 trafficked victims, preventing more slavery and illegal adoptions.

Ancient DNA and mitochondrial analysis

Microsatellites are extremely useful for analysing low-quality or degraded DNA left at a crime scene as their short sequences are usually preserved. However, DNA in specimens that have not been optimally preserved persists in exceedingly small amounts and is also highly fragmented. It is probably also riddled by contamination and chemical damage. Such sources of DNA are too degraded to obtain a profile using genomic STRs and in these cases mitochondrial DNA, being more abundant, is more useful than nuclear for DNA profiling. Mitochondria, the powerhouses of the vertebrate cell, are thought to descend from bacteria that moved into cells two billion years ago. The modest remaining genome of the mitochondria codes thirty-seven genes but there are many copies per cell compared to nuclear DNA. Thus there are hundreds of mitochondrial genomes to one nuclear genome. Mitochondrial DNA is inherited through the female line only and the mitochondrial DNA from all humanity has been traced back to a common female ancestor, dubbed 'mitochondrial Eve', who lived (among other women) in Africa some 170,000 years ago. Mitochondrial DNA profiling is the method of choice for determining the identities of missing or unknown people when a maternally linked relative can be found.

Molecular biologists can amplify hypervariable regions of mitochondrial DNA by PCR to obtain enough material for analysis. The DNA products are sequenced and single nucleotide differences are sought by comparison with a reference DNA from a maternal relative. An inconsistency of two or more nucleotides is sufficient to discount a match. Mitochondrial DNA testing was used to exclude the imposter Anna Anderson from being the Romanov Princess Anastasia.

The power of combining PCR amplification with DNA profiling opened the possibility of using DNA from skeletal remains and in 1989 this microsatellite technology was used to find the whereabouts of the notorious Dr Josef Mengele. The Auschwitz doctor had long been sought with the aim of obtaining justice for the victims of his war crimes. There were various alleged sightings, but other reports that he had died in Brazil. DNA profiles from his suspected remains in a Brazilian grave were compared with those of living relatives of Mengele and established the identity of the corpse as the infamous Auschwitz doctor.

The technology hit the headlines when it was applied to solving an eighty-year-old murder case in which the victim, unidentifiable at the time, was found in a burnt-out car. The psychotic killer Alfred Rouse killed the unidentified man and then placed the body in a car which he then set on fire. Relatives of a man who disappeared around this time wished to know whether the unknown victim was their ancestor. A small amount of tissue from the victim was available for a mitochondrial DNA comparison with DNA from the family to see if there was a match. The results excluded the man who had disappeared, and the identity of the victim of the Blazing Car Murder remains unsolved.

Next-generation sequencing: when a DNA match is not enough

No matter how powerful DNA fingerprinting has been it can only supply us with a matching identity. It would be extremely useful if crime-scene DNA could also reveal something about the physical appearance of the person from which a sample originated.

It has now become possible for crime-scene or ancient DNA to reveal much more than genotype matches. Microarray-based genotyping and GWAS technology has made it possible to identify genetic markers for complex traits, including what the person looked like. New advances in molecular biology, such as hybridization capture enrichment and next-generation sequencing, have radically improved our capability to analyse ancient DNA. Pigmentation characteristics can now be determined from ancient DNA since skin, hair, and eye colour are some of the easiest characteristics to predict. This is due to the limited number of base differences or SNPs required to explain most of the variability. Molecular biologists have also determined the candidate genes and molecular pathways that underpin these traits. The gene for melanocortin 1 receptor (*MC1R*) has been sequenced and found to have full or partial activity variants that are associated with darker or lighter pigmentation in humans respectively. Interestingly fully and partially active *MC1R* variants have also been found in Pleistocene mammoths, giving them dark and light hair colour. Some individual Neanderthals have been shown to have a largely inactive variant of the *MC1R* gene, which would have resulted in red hair and a fair skin that could be an adaptation to life in regions with low UV intensities where a pale skin would be a requirement for vitamin D synthesis. Very little other than pigmentation has been deduced about someone's appearance from sequencing their genes.

Ancient DNA and the Neanderthals

The remains of Neanderthals were discovered in Germany in the 19th century but were considered then as an evolutionary curiosity, unrelated to *Homo sapiens*. Neanderthals and modern man however did potentially overlap in Europe up to 30,000 years ago making interbreeding an intriguing possibility. Early attempts to answer the question of Neanderthal DNA in human genomes came back as negative but they were restricted to the use of mitochondrial DNA with its limited genome size. Recent genomic DNA sequencing has revealed that modern non-Africans are descended from ancient individuals who were the product of reproduction between humans and Neanderthal man. A comparison of genomic DNAs from Neanderthal and ancient hominin groups with non-African modern man demonstrates that our small inheritance of Neanderthal DNA is due to limited 'recent' interbreeding in geological terms. Therefore in the late Pleistocene period it appears that low levels of interbreeding did occur between Neanderthals and other hominin groups, including *Homo sapiens*. In particular modern man has inherited Neanderthal genetic information affecting keratin, the fibrous structural protein found in hair and skin that may have helped us to adapt to a non-African climate. Results suggest that part of the reason so little (only 1.5 to 2 per cent) of the Neanderthal genome now remains in modern man is that Neanderthal alleles or gene variants caused decreased fertility in males when moved to a human genetic background.

The DNA of Richard III identifies his remains

In 2012, skeletal remains were found under a car park in Leicestershire, England. Comparative mitochondrial DNA analysis from the skeleton and living relatives of Richard III as determined by genealogy gave a perfect match. For degraded ancient DNA, only mitochondrial DNA and the non-recombining part of the Y chromosome that is rich in repetitive sequences may be informative. As these sequences do not undergo recombination they are passed faithfully from father to son. For the mystery skeleton the Y chromosome sequences did not match present-day putative descendants, indicative of false paternity somewhere along the line. However maternal links are more reliable for obvious reasons and in this respect a good match was made between two descendants of Anne of York, Richard's sister, and the unnamed skeleton. Eye and hair colour were then typed from Richard's genomic DNA using probes to the relevant SNPs followed by direct PCR and sequencing for validation. This showed that Richard was probably a blue-eyed fair-haired boy although his hair colour would probably have darkened by adulthood to match that seen in his famous arched portrait. This genetic evidence, combined with physical features of the skeleton and carbon dating, forms overwhelming proof that Skeleton 1 found in a car park over the Grey Friars site in Leicester is that of the last Plantagenet king of England, Richard III.

Tackling pandemics and epidemics

Throughout history, cholera and other widespread diseases or pandemics have plagued the known world. Now the tools and techniques of modern molecular biology are helping us to understand how pathogens evolve, where pandemics come from, and what if anything, can be done to stop them. As bugs and hosts co-evolve, strains of the pathogens arise that are more suited to the host and therefore less lethal. This has important consequences for cholera outbreaks.

Cholera is caused by several different strains of a bacterium called *Vibrio cholerae* and still causes 58,000–130,000 deaths a year worldwide. Dr Hendrik Poinar working at McMaster's University wanted to study evolution of cholera strains. 'The best way to do that', he said, 'is to focus on the past, to decode ancient pathogens and compare them to their modern counterparts to see how they have changed over time'. Around 1850 in Philadelphia, a man succumbed to cholera. This was not surprising as the world was in the grip of the second severe cholera outbreak, but what was unusual for that time is that a piece of his infected intestine was preserved in alcohol. One hundred and sixty-five years later DNA was extracted from this ancient tissue but unfortunately the nucleic acid was extremely degraded, containing fragments that averaged only thirty-five base pairs in length. The total DNA was however sequenced and the results showed that the DNA was not only from the *Vibrio cholerae* bacterium and human genomic sequences as might be expected but also from the bovine tissue from a previous use of the pickling jar.

This illustrates the problems facing work on ancient DNA samples. Using a DNA microarray carrying oligonucleotides complementary to the cholera genome, as well as human mitochondrial DNA, the *Vibrio* DNA sequences were then purified, sequenced, and results assembled. These sequences were then compared to those from known cholera genomes from two recent outbreaks in the early and mid-20th century, known as the sixth and seventh pandemics respectively. The second pandemic from the 19th century was caused by the classical cholera strain that plagued the world up to the sixth pandemic (1899–1923). However this classical cholera has now disappeared from the earth and been replaced in the latest or seventh pandemic by the El Tor strain. So for over 150 years the same cholera strain caused repeated waves of infection but it has now been replaced by a strain that is less destructive but still potentially lethal.

Molecular biology is now at the heart of combating a modern epidemic, this time viral: that of Ebola disease. The Ebola virus was discovered in 1976 and is part of the group of viral haemorrhagic fevers. The virus genome is a single strand of RNA,

19kb long, that is coated in protein and lipid to form the virion. One of these proteins is the viral-RNA-dependent RNA polymerase required for transcription of its seven genes. The very cells that usually counteract new invaders in the body, the macrophages, monocytes, and dendritic cells, are the preferred sites of replication for the virus. These infected cells then take the virus around the body. The virus alters the immune system by inducing the expression of pro-inflammatory molecules such as interferon and interleukins that actually have a detrimental effect on the sufferer. These small proteins or cytokines make blood vessels leaky and initiate coagulation of the blood. This widespread clotting within blood vessels uses up all the body's clotting factors leading to internal bleeding, major organ failure, and death.

A reservoir of virus is thought to exist in monkeys and bats, which presents a risk for further outbreaks if eaten or just from faecal contamination. Early detection of new cases is vital to prevent rapid development of further epidemics like the one in Sierra Leone. The major diagnostic test at present is an RT-PCR for its RNA genome but this requires the high level of biosecurity precautions usually only available in specialized laboratories. ELISA testing—a method which uses antibodies to detect the protein coat of the Ebola virus—is also available but a positive test will only be obtained once the viral load is high enough for the patient to be showing symptoms.

A barcode for a species

A species is a group of organisms capable of mating and producing fertile offspring, but in the absence of a breeding experiment how can we be sure of identifying a species? Paul Hebert working at the Biodiversity Institute of Ontario came up with the idea of DNA barcoding as a means of answering this question. DNA barcoding uses a DNA sequence within a single gene region that is found in a wide range of organisms that lends itself to standardization across species.

Molecular biologists have found that within the 37-gene mitochondrial genome resides a 648-nucleotide sequence that is a powerful tool for distinguishing species. This mitochondrial barcode varies very little within a species but has little or no overlap between species. It varies at no more than two positions between any two humans but between the chimpanzee, our nearest relative, and us it differs at sixty sites. The barcode sequence allows us to identify unknown species and classify new ones. For example, ten separate species of the blue-flasher butterfly have now been identified since the barcode enables us to distinguish between closely related species. Libraries that can be used for comparison and reference have been generated using this mitochondrial DNA fragment from the cytochrome oxidase C gene. This sequence was chosen because it works better than anything else. Why does it work so well? It could be that after two species split from a common ancestor the sequences soon change to become unique. This is plausible since mitochondrial DNA mutates ten to thirty times faster than nuclear DNA but it does not really explain why DNA barcodes vary remarkably little within a species. An additional explanation would be that rather than being a bystander this mitochondrial sequence might actually be a powerful driver in the process of speciation. The mitochondrion is the powerhouse of the cell, releasing energy from 'burning' food by respiration. Not all the genetic information for making a mitochondrion resides within mitochondrial DNA; some genes are also located within the nuclear genome. The enzyme cytochrome oxidase C from which the 'barcode' is derived is part of a complex that catalyses the last step in cell respiration. Nuclear genes encode several of the subunits that comprise this enzyme complex. The two genomes, mitochondrial and nuclear, must work in harmony therefore for the subunits of cytochrome oxidase to pass electrons to oxygen for successful respiration. The penalty of failure is death.

When environmental conditions change and a new food source becomes available organisms must select new enzymes to avail themselves of it. Mitochondrial genes by mutating rapidly generate the variation necessary for this selection. However mutations in the nuclear genome must be selected to remain in functional harmony with the mitochondrial DNA and only organisms that can ensure that their mitochondrial and nuclear genomes function together will survive. Mitochondrial genes only come from the mother but if they do not function in the nuclear background of the new offspring there is a dangerous decrease in fitness, referred to as hybrid breakdown. Thus changes in the mitochondrial genome, to allow say a new food source to be used for energy, would require the selection of suitable changes in the nuclear genome in order to maintain function. When this adapted individual then mated with an individual from the original population an imbalance between the two genomes could lead to non-viable offspring. This failure to mate successfully signifies the two individuals are from separate species. Thus the DNA barcode may well succeed at tracking species because it is closely linked to creating them.

Biosecurity—using mitochondrial DNA

Biosecurity encompasses, among other aspects, protecting borders from unwanted importation of harmful pests requiring accurate and swift identification of morphologically indistinct alien species. This is particularly important for identifying immature life stages of invertebrates that could have profound impacts on ecosystems or the economy of a country heavily dependent on agriculture. DNA barcodes can provide a valuable tool for species identification in the context of biosecurity. In 1999 a devastating new arrival in the form of the painted apple moth arrived in New Zealand. Although a minor inconvenience in its native Australia it was predicted to have a significant impact on the fruit-growing industry in New Zealand of up to 200,000,000 NZ$ over the next twenty years if not eradicated. DNA barcoding was invaluable as a tool for identifying the pest and tracking its whereabouts during the eradication programme. Fruit flies, intercepted at the New

Zealand border, have also been identified by DNA barcoding. Previous methods relied on fresh tissue suitable for immunological or protein-based assays. DNA-based assays, particularly in combination with PCR for sensitivity, are more suited to biosecurity issues and pursuing eradication where necessary.

How can we identify food counterfeits?

A number of scandals relating to species counterfeit and mislabelling of food products have entered the media. A study conducted by the company Oceana in 2012 reported that one-third of seafood samples taken from 673 retail locations across the USA were of a different species from the one specified to the consumer. Once a fish is filleted it is difficult to identify. Hence cheap alternatives can be passed off as luxury species or endangered species can be sold beyond quota. Whether mislabelling is intentional or inadvertent, it can pose significant health risks to the public. In 2007, toxic puffer fish were sold as monkfish causing severe sickness in consumers. Escolar, an oily fish, has also been passed off as white tuna. Escolar is not a tuna species but a snake mackerel species that contains a naturally occurring toxin. For those who eat too much of it, severe gastrointestinal problems can occur. More recently, a seafood company based in Florida was charged and fined for importing Chilean steelhead trout and selling the fish relabelled as the more expensive salmon. Previously protein extracts were analysed to identify seafood species but this was far from satisfactory. Now DNA barcodes are providing a powerful means of investigating seafood fraud. The Canadian Barcode of Life Database (BOLD) includes DNA barcodes for over 8,000 species of fish that can be used for quick identification.

DNA profiling was also used to identify horsemeat in beef products in the scandal that started in the UK and Ireland and evolved into a pan-European incident. It happened when the

Food Safety Authority of Ireland conducted tests on products labelled as '100% beef'. They discovered horsemeat in cheaper hamburgers and in some frozen ready meals including beef lasagne and spaghetti bolognaise. A PCR-based DNA fingerprinting method was used to differentiate horsemeat from beef. The findings led to millions of beef burgers, ready meals, and mince packs being withdrawn from supermarkets and restaurants. Since then, retailers have been working hard to regain the trust of the consumers by reassessing their sourcing and increasing testing of ingredients.

Chapter 9
Future challenges

Advances in molecular biology provide us with opportunities for tackling chronic global issues. Two areas of molecular biology that will have significant impact on society in the coming years are non-communicable diseases and synthetic biology.

Molecular biology is being applied to some of the most serious challenges facing our future health and longevity. These are the obesity epidemic, personalized medicine, and the impact of genetic diseases. Diet and lifestyle are implicated in two-thirds of age-related diseases such as diabetes, cancer, and cardiovascular disorders. The morbidity and economic consequences of the obesity epidemic is of great concern in an ageing population and at present the cost of obesity-related type II diabetes alone accounts for over 1 per cent of worldwide gross domestic product. Also of relevance are delayed parenting and the increase in assisted fertility and again molecular biology is at the forefront of new technologies in this field. As we have seen, cancer is not a single disease but comes in hundreds of different forms, depending not only on the site and cell of origin but more importantly on the spectrum of genomic alterations that promote its formation. All these factors affect a therapeutic response. It is becoming increasingly obvious therefore that most cancers will require diagnosis at a molecular level so that personalized medicine can be offered. Personalized medicine refers to the tailoring of treatment to a patient, based on the specific molecular makeup of their disease. Genomic and proteomic technologies have revolutionized our understanding of disease processes and we are now translating this knowledge into a clinical setting for diagnosis and treatment.

'First do no harm' is one of the fundamental principles of medical ethics and this points to a personalized approach to treatment. Until recently, individuals with the same disease will have received the same treatment, with mixed results. Some respond well to these treatments while others do not respond or even experience serious side effects. In personalized medicine, the approach is to identify biomarkers—DNA, RNA, or protein molecules that are specific to a particular disease or disease stage. Personalized medicine is an evolving area of research with huge amounts of resources dedicated to searching for biomarkers that can be accurately measured to guide diagnosis and treatment. One area in which biomarkers are currently being used in the clinic is to treat cancer. High-throughput molecular techniques such as next-generation sequencing of a patient's cancer can be used to identify the specific molecular defects it harbours. Finding the molecular Achilles' heel of a person's cancer is the first step in curing it. We talk of 'addiction' of a tumour to a molecular defect that the tumour relies on for its survival. Blocking or removal of this defect is key to killing that cancer.

Biomarkers for therapeutic prediction in cancer

A well-known example of personalized cancer treatment is the use of the humanized monoclonal antibody trastuzumab, commonly known as Herceptin. This drug is used to treat breast cancer patients whose tumours overexpress the oncogene HER2. HER2 testing is performed on breast cancer biopsy tissue to determine suitability for treatment with Herceptin. HER2 is a growth factor receptor overexpressed in certain types of breast cancer that drives proliferation and aggressive behaviour. It is an important biomarker for targeted therapy in 15–30 per cent of breast and other types of cancers. These tumours are addicted to the deregulated receptor and thus blocking it with Herceptin halts tumour proliferation. Only patients with HER2 positive tumours will benefit from Herceptin so it is important that the drug is restricted to these patients.

The challenge for the future is to characterize new oncogene addictions, provide relevant molecular targets for drug development, and to document the clinical benefit of their blockade. Relentless progress in next-generation sequencing from biopsy-derived molecules is promoting a wave of new genomic, transcriptomic, epigenomic, and proteomic technologies to provide this information. There is a note of caution, however, that some changes assumed to be tumour specific may be present in normal cells. This requires sequencing-based technologies that are automated to enable high-throughput and multi-dimensional analyses of individual cells.

The desire for less invasive methods than tissue biopsy has led researchers to actively study blood-based biomarkers for cancer and other diseases, and microRNAs (miRNAs) are possible new candidates. These small regulatory RNA molecules are implicated in many diseases—cancer, neurological disorders, and cardiovascular disease among others. They circulate in body fluids in a stable condition making them potentially powerful biomarkers but new technologies are required for their accurate measurement. The answer may lie in digital PCR that until recently has been hampered by high costs. Digital PCR is a variant of the conventional method that enables rare variants to be measured accurately in a background of wild type molecules. Digital PCR also enables white blood cells that infiltrate the tumour to be counted accurately so that patients can be stratified into groups for treatment with new immunotherapeutics. These are drugs designed to block cancer cells from evading the immune system, thus allowing natural immunity to eliminate them.

Mitochondrial DNA and heritable diseases

A broad range of debilitating and fatal conditions, none of which can be cured, are associated with mitochondrial DNA mutations. Since mitochondrial DNA is maternally inherited, if a woman carries a DNA mutation in her mitochondria, she is at risk of passing this on to her children. Mitochondrial DNA mutates at a higher rate than nuclear DNA due to higher numbers of DNA molecules and reduced efficiency in controlling DNA replication errors. Mutations in mitochondrial DNA can cause both miscarriage and disease in up to 1 in 200 live births in the UK. Mitochondrial diseases range in severity from debilitating to fatal. Since mitochondria provide most of the energy in a cell, deleterious mutations are felt most acutely in energy-demanding tissues such as the heart, brain, and skeletal muscle. For example, some forms of muscular dystrophy are due to mutation in mitochondrial DNA. More western women are postponing parenthood, and this can carry an increased risk of infertility, partly due to aged and mutant mitochondria. Correcting these mitochondrial defects may reduce infertility and in vitro fertilization (IVF) failures.

Over 100,000 copies of mitochondrial DNA are present in the cytoplasm of the human egg or oocyte. After fertilization, only maternal mitochondria survive; the small numbers of the father's mitochondria in the zygote are targeted for destruction. Thus all mitochondrial DNA for all cell types in the resulting embryo is maternal-derived. Usually not all the mother's mitochondria carry any specific mutation, a condition known as heteroplasty. Patients affected by mitochondrial disease usually have a mixture of wild type (normal) and mutant mitochondrial DNA and the disease severity depends on the ratio of the two. Importantly the actual level of mutant DNA in a mother's heteroplasty is not inherited and offspring can be better or worse off than the mother. This also causes uncertainty since the ratio of wild type to mutant mitochondria may change during development. This process is currently poorly understood but an increase in the proportion of mutant DNA gives rise to progressive disease which is usually more obvious in non-dividing tissues such as brain and heart giving rise to cognitive dysfunction and myopathies.

An example of a mitochondrial disease is Pearson's syndrome, in which a 5-kb deletion of mitochondrial DNA results in anaemia in infancy. Children who survive to adolescence develop a myopathy as the mutant DNA increases in their muscles. Over 700 mutations in mitochondrial DNA have been found leading to myopathies, neurodegeneration, diabetes, cancer, and infertility. Intervention to prevent these devastating diseases is obviously an attractive prospect and mitochondrial replacement (MR) using three-way IVF could provide the answer. A female donor provides the mitochondrial DNA with the nuclear DNA coming from the mother and father, effectively giving the embryo three parents (see Figure 24).

There are concerns about the technique since we are still ignorant of the consequences that the mitochondrial genome might have on the offspring, either directly through its role in providing cellular energy, or indirectly by moderating activities of the nuclear genome. Molecular biology has shown that the mitochondrial genome has evolved to be in harmony with its nuclear counterpart and the effects of disrupting this balance may be severe. Recent studies in mouse models suggest that using mismatched mitochondria and nuclear genomes during MR could trigger an immune response or lead to physiological and behavioural changes in the offspring. Mitochondria certainly have functions well beyond simple energy production, reinforcing concerns about MR.

Donor and recipient may need to be matched similar to blood transfusions in order to prevent reduced fitness as a consequence of the breakdown in co-evolved mitochondrial DNA–nuclear

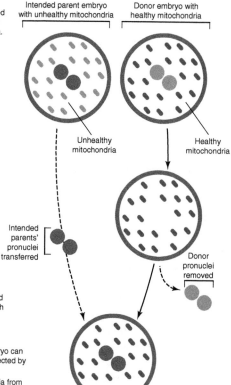

Pronuclear transfer (PNT)

Step 1
The intended parents' sperm and egg are used to form an embryo with unhealthy mitochondria. A donor embryo with healthy mitochondria is also formed.

Intended parent embryo with unhealthy mitochondria

Donor embryo with healthy mitochondria

Unhealthy mitochondria

Healthy mitochondria

Step 2
Donor pronuclei are removed from the embryo with healthy mitochondria.

Intended parents' pronuclei transferred

Donor pronuclei removed

Step 3
The intended parents' pronuclei are transferred to the donor embryo with healthy mitochondria.

Outcome
The reconstructed embryo can go on to develop, unaffected by mitochondrial disease.

The healthy mitochondria from the donor will now be passed on down the maternal line through future generations.

Future challenges

24. Nuclear DNA from the mother's egg is transferred to a healthy donor egg cell that lacks its nucleus. The mother's DNA is held in chromosomes in a structure called the spindle. This technique is therefore known as 'spindle-chromosomal complex transfer'.

interactions. The effects and viable thresholds of mismatches in mitochondrial and nuclear DNAs are required, as are measurements of maternal mutant mitochondrial DNA carryover to the offspring. Work is required to characterize these phenomena and provide evidence that would reassure potential parents that they do not impose sufficient threat to prevent the application of mitochondrial replacement therapy.

Synthetic biology

Synthetic biology is another exciting and evolving area of research. It combines genetic engineering techniques with the practical principles of engineering to design and build new biological systems or organisms with novel or enhanced properties. As natural systems are highly unpredictable, the aim of synthetic biology is to make these more reliable and in doing so extend the scope of biological functions. This in turn can be used to improve human health and address environmental and agricultural issues. In engineering, a system is built from devices that are composed of individual parts. For example, a car (the system) is a multi-component device (including engine, chassis, and gearbox), each derived from individual parts (tyres, brakes, clutch, and so forth). This approach of parts, devices, and systems is also used in synthetic biology. The parts, called BioBricks, are pieces of DNA that encode a biological function. Among these are gene coding sequences, promoter and terminator sequences, and ribosomal binding sites. Devices are built from a collection of bioparts and a system generated from combining the devices (see Figure 25). Systems are designed to carry out a specific task—for example, a biosensor (the system) for diagnosing disease, or a cell (the system) for synthesizing a particular protein. In synthetic biology, as in engineering, the starting point is to define the specification of the part, device, or system and then develop a design that meets these requirements. This process is cyclic, requiring repeated rounds of detailed computational modelling and practical testing, with the product

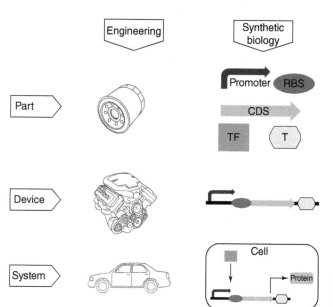

25. Synthetic biology uses a parts, devices, and systems approach, similar to engineering. CDS = coding sequence, RBS = ribosomal binding site, T = terminator, TF = transcription factor.

being refined with each cycle. This approach ensures that regardless of who manufactures the device or part, all are constructed to a particular specification. The characterized or standard building blocks are stored in databases and are freely available to other synthetic biologists.

One of the earliest successes of synthetic biology was the production of semi-synthetic artemisinin, an anti-malarial drug that is naturally produced by the sweet wormwood plant *Artemisia annua*. This plant has been used in traditional Chinese medicine for many years and in the 1970s, Chinese scientists identified artemisinin as the anti-malarial ingredient. Malaria is a disease affecting 300 million people each year and is particularly common in Africa.

Almost thirty years later, the WHO recognized artemisinin-based drugs as an effective treatment against the malaria-causing parasite *Plasmodium falciparum*. As artemisinin is extracted from plants, the drug supply and pricing can fluctuate widely as plant yield varies with climatic changes. To ensure an affordable and reliable source, the Semi-synthetic Artemisinin Project was launched in 2004 funded from the Bill and Melinda Gates Foundation. The aim was to engineer a micro-organism to produce an artemisinin precursor at high levels which could then be extracted and converted to the actual drug using industrial processes. To do this, synthetic biologists constructed a metabolic pathway involving a number of sequential steps within the host organism yeast to produce the precursor. Large-scale production of the anti-malarial drug was achieved in 2013 and, a year later, the drug manufacturer Sanofi released its first batch of semi-synthetic artemisinin.

Integrating defined or reconfigured biochemical pathways into existing host organisms is one aspect of synthetic biology. Another more controversial area is the construction of new cells carrying entirely synthetic DNA. Built from basic chemical and biochemical building blocks containing the minimum number of genes, these minimal cells would be engineered to act as specialized hosts, or chassis, each carrying out a specific function. The advantage would be that the biological functions of all the genes added would be known and the inherent unpredictability of natural cells could be overcome. The first chemically synthesized cell genome was pioneered by researchers at the J Craig Venter Institute and the work published in 2010. They assembled a complete 1 million base pair circular genome of the bacterium *Mycoplasma mycoides* from synthesized DNA fragments. The genome was assembled in yeast and then transplanted into a bacterial cell that was empty of DNA. The cell is referred to as a 'synthetic cell', even though only the genome is chemically synthesized, not the recipient cell.

Synthetic biology has the potential to provide society with more effective medicines, cheap biofuels, and re-engineered crops with increased nutrients and enhanced yield. However, this advancing technology raises ethical and social concerns about the risks to human health, environmental contamination, and deliberate misuse which are concurrently being considered.

Genome editing

In Chapter 6 we described processes whereby a recombinant DNA construct is introduced into mammalian cells for a number of purposes, such as to produce recombinant protein, to generate transgenic organisms, and for gene therapy. Typically these DNA constructs integrate randomly into the host genome, causing unwanted effects. However, a new technology—genome editing—is changing this, allowing researchers to disrupt gene sequences or add new sequences with pinpoint precision. This technology uses artificially engineered endonucleases or 'molecular scissors' as they are called to create double-stranded DNA breaks (DSBs) at specific sites within the genome. This activates the cells' natural repair mechanisms of which there are two types, homologous recombination (HR) and non-homologous end joining (NHEJ). In HR the damage is repaired using an identical or homologous copy of the broken chromosome as a template and is essentially an error-free repair process. In genome editing, endonucleases are co-delivered with the desired DNA sequence, which is then used as a template by the HR system, thus introducing the desired change of sequence at the location of the break. In NHEJ, the broken chromosome ends are simply joined back together without using a repair template. This is an error-prone process frequently leading to small insertions or deletions termed Indels, at the break site. In genome editing, NHEJ is used to knock out the expression of a gene as Indels often cause loss of protein expression or function (see Figure 26).

In the last ten years, four classes of genome-editing nucleases have been designed. Of these nucleases, CRISPR-Cas9, is revolutionizing molecular biology and has been described 'as the

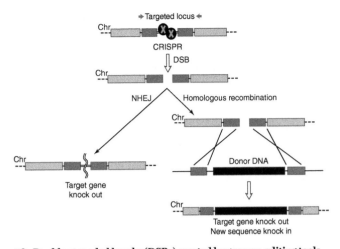

26. Double-stranded breaks (DSBs) created by genome-editing tools, trigger the DSB repair mechanisms within the cell. Left: Non-homologous end joining. Right: Homologous recombination in the presence of a donor template.

biggest game-changer to hit biology since PCR'. CRISPR is an acronym for clustered, regularly interspaced, palindromic repeats and is composed of an RNA strand designed to home in on its target DNA coupled to the DNA cutting enzyme Cas9. DNA breaks generated by Cas9 at target locations are repaired by NHEJ or HR. Since the first publication in 2012 showing that CRISPR-Cas9 could be used to edit genomes, it has been widely adopted by researchers, facilitated by the ease and low cost with which it can be engineered compared to the earlier genome-editing nucleases.

CRISPR can be used in virtually any organism or cell type—microbes, plants, animals, and humans—to cut out unwanted DNA sequences, or add a new stretch of DNA into a selected location. It is being applied to edit and assemble complex metabolic pathways in synthetic biology, to knock out or knock in genes for disease modelling, and to transfer specific genetic traits in agricultural crops and livestock. Genome-editing technology also has the potential to accelerate the treatment of genetic diseases. The first use of CRISPR to correct a genetic disease in an adult animal was published in 2014. Tyrosinaemia, a human inherited disorder, is caused by a point mutation in the FAH gene. FAH codes for an enzyme responsible for breaking down the amino acid tyrosine, a building block of most proteins. The mutation results in the absence of a functional enzyme leading to the accumulation of toxic products and causing liver damage. Researchers Hao Yin and colleagues showed that CRISPR-Cas9 could correct the FAH mutation and restore functional enzyme production. This work was carried out in a mouse model of the human disease and its success brings researchers a step closer towards using genome-editing therapy in humans. Currently, genome editing is being investigated for application to a number of genetic diseases including haemophilia B, cystic fibrosis, and Duchenne muscular dystrophy. It is also being studied as a possible means of treating viral diseases, such as HIV and hepatitis B, by silencing critical elements of the viral genome. However, the technology is still in its infancy and there is much that needs to be known before it can be used safely and efficiently in the clinic.

The future potential of these molecular biology techniques is enormous, and if channelled appropriately, they have the power to improve health and benefit the environment.

References

Chapter 4: Proteins

1. Marx, V. (2014). Proteomics: an atlas of expression. *Nature*, 509: 645–9.
2. Khoury, M. P. & Bourdon, J. C. (2010). The isoforms of the p53 protein. *Cold Spring Harbour Perspectives in Biology*, 2(3): a000927.

Chapter 5: Molecular interactions

1. Dias, B. G. & Ressler, K. J. (2014). Parental olfactory experience influences behavior and neural structure in subsequent generations. *Nature Neuroscience*, 17(1): 89–96.
2. Kerr, J. F. R., Wylie, A. H., and Currie, A. R. (1972). Apoptosis: a basic biological phenomenon with wide-ranging implications in tissue kinetics. *British Journal of Cancer*, 26: 239–57.

Further reading

Textbooks

Bruce Alberts and Alexander Johnson, *Molecular Biology of the Cell* (Garland Science, 2014).

Jocelyn E. Krebs, Benjamin Lewin, Elliott S. Goldstein, and Stephen T. Kilpatrick, *Lewin's Essential Genes* (Jones & Bartlett, 2013).

Lauren Pecorino, *Molecular Biology of Cancer: Mechanisms, Targets and Therapeutics* (Oxford University Press, 2012).

Reading books

Sue Armstrong, *p53: The Gene that Cracked the Cancer Code* (Bloomsbury Sigma, 2014).

Nessa Carey, *The Epigenetics Revolution: How Modern Biology is Rewriting our Understanding of Genetics, Disease and Inheritance* (Icon Books, 2012).

John Parrington, *The Deeper Genome* (Oxford University Press, 2015).

James D. Watson, *The Double Helix* (Weidenfeld and Nicolson, 1968).

Academic articles

D. Baltimore, 'Our genome unveiled', *Nature*, 409(6822) (2001), 814–16. This article discusses the landmark publication of the draft human genome sequence and what it means.

D. E. Cameron et al., 'A brief history of synthetic biology', *Nature Reviews Microbiology*, 12(5) (2014), 381–90. This timeline article charts the technological and cultural lifetime of synthetic biology, with an emphasis on key breakthroughs and future challenges.

L. Chin et al., 'Cancer genomics: from discovery science to personalized medicine', *Nature Medicine*, 17(3) (2011), 297–303. This article provides a perspective on some of the major obstacles to translating cancer genome discoveries to a clinical setting (bench to bedside).

J. R. Ecker et al., 'Genomics: ENCODE explained', *Nature*, 489(7414) (2012), 52–5. This article summarizes the findings from the landmark ENCODE project.

T. King et al., 'Identification of the remains of King Richard III', *Nature Communications*, 5 (2014), 5631.

H. Ledford, 'End of cancer-genome project prompts rethink', *Nature*, 517(7533) (2015), 128–9. In this news article, scientists debate whether the focus should shift from sequencing genomes to analysing function.

P. Sahu et al., 'Molecular farming: a biotechnological approach in agriculture for production of useful metabolites', *International Journal of Research in Biotechnology and Biochemistry*, 4(2) (2014), 23–30. This article summarizes the production of pharmaceutical drugs, including vaccines and therapeutic proteins in plants, and discusses associated issues.

Web resources

CRISPR: the good, the bad and the unknown. In this special issue *Nature* brings together research, reporting, and expert opinion on gene-editing and its implications (2015). Available at: <http://www.nature.com/news/crispr-1.17547>.

GM crops: promise and reality. In this special issue *Nature* explores the hopes, the fears, the reality, and the future of GM crops (2013). Available at: <http://www.nature.com/news/specials/gmcrops/index.html>.

Scitable by Nature Education. This is an educational resource covering lots of different life science topics from the *Nature* Publishing Group. Available at: <http://www.nature.com/scitable>.

Index

Index

Molecular Biology

Index